A Garden of Marvels

ALSO BY RUTH KASSINGER

Paradise Under Glass

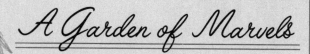

A Garden of Marvels

How We Discovered That Flowers
Have Sex, Leaves Eat Air, and
Other Secrets of Plants

RUTH KASSINGER

wm
WILLIAM MORROW
An Imprint of HarperCollins*Publishers*

HarperCollins books may be purchased for educational, business, or sales promotional use. For information please e-mail the Special Markets Department at SPsales@harpercollins.com.

FIRST EDITION

Designed by Lisa Stokes

Library of Congress Cataloging-in-Publication Data has been applied for.

ISBN 978-0-06-204899-8

14 15 16 17 18   OV/RRD   10 9 8 7 6 5 4 3 2 1

To Kenneth Greif, mentor of a lifetime

Contents

Introduction

This book was born of a murder, a murder I committed. It was not my first, but I have some hope it will be my last. Since I never set out to kill—quite the contrary—I suppose I am guilty only of negligent homicide, or possibly mere criminal negligence. Still, I feel deeply culpable. All I can do is plead ignorance, and say that this particular death was a life-changing event for me (as well, of course, for my victim). Possibly, since you have this book in your hands, the tragedy will save a few lives I will never know.

The deceased in this case was a twelve-year-old guest, a permanent resident, really, of my household. She was a lovely, graceful creature about five feet tall, and a particular favorite of my family. Kam Kwat she would have been called in Cantonese, had she lived in her native land. As it was, since we live just outside Washington, D.C., we knew her as the kumquat tree. In the summer she lived outdoors on our patio in a cobalt-blue pot; from late fall through winter and into spring she lodged in the small conservatory attached to the side of our house. I carried out the murder—and I can't help feeling her death deserves that name—in the conservatory, with the clippers.

The facts are as follows. I brought my kumquat tree home as a three-year-old, an immigrant from a nursery nearby. She was a slender thing, and her gracile limbs, covered in deep green, oval leaves, lifted upward and curved slightly outward, as if their weight were just a bit more than her youthful frame could bear. The following spring she blossomed, flushing with tiny, white, and fragrant flowers. In a week or so, the flowers fell, revealing dark green beads at their bases. These gradually expanded and finally ripened into golden-orange fruits the size of large grapes. Plucked off the tree and eaten in one bite, the fruit had a slight bitterness redeemed by the sweetness of the delicate peel. Even our terrier liked them, and ate any that dropped to the floor.

Every year for five years, the tree grew a few inches taller, and her limbs branched again and again, until she wore a dense and glossy crown. At nine, though, she stopped growing. I was relieved, actually, since the conservatory is crowded in winter after I bring in the orange, lemon, fig, coffee, bay laurel, loquat, and banana trees, and a host of potted herbaceous plants. The following winter, however, she began to decline. The lower leaves on her branches turned yellow and, one after another, fell to the stone floor with a delicate rattle. By early spring when I moved her outside again, she held skinny, bare arms to the sky; only their very ends were leafy, as if she were wearing nothing but green gloves. Was this a watering problem? I'm a lackadaisical gardener, so I tried to be more conscientious with my watering can. Was she hungry? I sprinkled her soil with fertilizer pellets. I repotted her into a larger container.

Nothing helped. I looked to Ted, my husband and a more accomplished gardener, for inspiration. We have a strict division of labor in our household: I take care of the indoor plants and Ted tends to the outdoor ones. Years ago, he planted a crape myrtle just outside a living room window, and early every spring, before its new leaves appear, he prunes its multiple trunks back to windowsill height. About the same time, he razes the hedge of Knock Out rosebushes that line our driveway, reducing them to a two-foot-high stubble. Miraculously (to my mind), within weeks all the plants leaf out. By mid-May, we have rosebushes hectic with red blooms. In July, the crape myrtle is weighted with magenta flowers that fill the window frame, just where we can enjoy them. There seemed to be a lesson in this for me. I broke out the clippers and gave my kumquat a thorough—really thorough—pruning. Then I put her outside in the sun, and waited.

Her branches turned brown and brittle. I had killed her.

Why? Why did she succumb when the crape myrtle thrived? It occurred to me I had run up against some fundamental biological difference between crape myrtles and kumquat trees, but I had no idea what it was.

This failure brought to mind other local botanical mysteries. Some friends up the street had a beautiful hickory. During a construction project, they added a deep layer of soil to their yard, and although they constructed a tree well around the trunk, the tree died. Why? Why does the vibrant green fescue in our lawn inevitably lose ground to scraggly, tough crabgrass? Why do houseplants, so beautiful and vigorous in the garden center, decline after a few weeks at home, even if I tend to them like a robin doting on her newly hatched chicks? Most immediately, I had two tree-size dracaena, those foolproof foliage plants ubiquitous in offices and shopping malls, whose roots had grown through the drainage holes in their pots and now circled their saucers. Could I safely prune these ugly outliers? I knew the basic rules of caring for indoor plants—drain any water in plant saucers, let the top inch of soil go dry before watering, apply fertilizer in spring—but I realized I understood very little about how roots, stems, leaves, and flowers function, what physical and chemical processes make them work. In other words, I knew almost nothing about the physiology of plants.

How could it be that I knew so little? Here's one excuse: I'm a suburbanite, and live in a house with a yard the size of a carpet remnant. Yes, where I live oak trees drop acorns, daffodils bloom, and grass (or crabgrass) grows. But vegetation here is a minor component of a landscape dominated by man-made features like houses, streets, and cars. Until I took up indoor gardening, the only thing I pondered about plants was which ones to eat.

Another excuse: I'm a humanities person, and went through college and graduate school without taking a single science course. I was so inattentive in my high school science classes—I was to be a poet, you see—that I was unprepared for even the basic college-level biology course. I came to regret those academic choices, and over the years have learned enough physics and chemistry to write books for young adults on the science and history of inventions and materials. But of plant science, I knew almost nothing. Here's yet another excuse: High school biology textbooks then (and now) usually devote only a chapter or two to plants, and largely as a warm-up to what is the presumably more interesting subject of animal biology.

A few botanical facts did reverberate faintly in my brain, like the half-remembered lyrics of pop songs I listened to in the same era. The words *xylem* and *phloem* I recalled because they regularly appear as answers to clues in crossword puzzles. I even knew that both are tubes in a plant for transporting water and . . . and . . . well, and something else. *Pistils* and *stamens* are the male and female parts inside a flower, although which was which I had forgotten. Photosynthesis, I knew, involves taking carbon dioxide out of the air while simultaneously producing oxygen. How that transformation occurs, I couldn't say. I must have daydreamed through Mrs. Miller's explanation in ninth-grade biology.

When in middle age I began caring for a collection of indoor plants, I turned to gardening books and hang tags for the practical advice I needed. I was moderately successful gardening in this way, although, as I say, Kam Kwat was far from my first failure. It now dawned on me that blindly following instructions—or acting on my own mistaken analogies—without any insight into the science of plants might be part of my problem. Maybe if I knew how a kumquat differs, physiologically, from a crape myrtle, I might avoid another murder. I started off to educate myself, hoping to become a better, or at least a less lethal, gardener.

I began by checking out botany and plant physiology text-books from the University of Maryland library, and set out to work my way through them. To my chagrin, I made little progress. There was too much information, too much detail. Some was pertinent to my interest as a gardener, some was fascinating but irrelevant, too much was inscrutable. Botany felt like a steep and thorny cliff that I was trying to summit by climbing straight up its vertical face. What I needed was to circle around to the far side and find a gently sloping path to the top. Let there be some curious plants along the way, intriguing historical markers from time to time, and the occasional overlook where I could get a long-range perspective. Maybe I would come across some fellow hikers to chat with as I ambled along.

My primrose path, I decided, would be the story of the first discoverers of the basic facts of plant life. I would go back to the beginning and retrace, step by step and insight by insight, their progress in understanding the way plants work. When I looked for a history of plant physiology, however (as opposed to plant classification), I found little. The only modern chronicle is *History of Botanical Science: An Account of the Development of Botany from Ancient Times to the Present Day* (Academic Press, 1981), by Professor A. G. Morton at the University of London, and he wrote it for his botany students. It is an excellent book, but hardly the saunter I had in mind. Botany 101 is definitely a prerequisite, and the human side of the story was of little interest to the professor. If he was a gardener, his interest is undetectable. Still, his book made me think that history would be the route I needed.

It turns out that these early explorers of the vegetal world were a diverse, quirky, and unfairly neglected bunch. The men—and they were all men, since we're talking about the late seventeenth and eighteenth centuries—include a melancholy Italian anatomist, a renegade French surgeon, a stuttering English minister, an obsessive German schoolteacher, and Charles Darwin, who, in the last two decades of his life, devoted most of his intellectual energies to botany.

Intrepid though they were, they uncovered the fundamental facts of botanical science remarkably late, compared to other sciences. In 1670, no one yet knew what lies inside a stem, much less how water moves up a trunk, or what plants eat to grow. Consider that by that date, the anatomist Andreas Vesalius had published accurate descriptions of the human skeleton, musculature, and many internal organs. William Harvey had revealed the basics of the human circulatory system and Thomas Bartholin had discovered the human lymphatic system. Physics was far advanced. Johannes Kepler had developed mathematical laws that define planets' orbits; Galileo had explained how the Earth revolves around the sun and had seen the moons of Jupiter; and Isaac Newton had invented calculus, discovered the law of universal gravitation, and revealed that sunlight is composed of all the colors of the spectrum. Evangelista Torricelli had invented the mercury barometer and created an artificial vacuum, and Christiaan Huygens had constructed a pendulum clock. Yet no one knew what flowers are for.

There are reasons for botany's tardiness. Observe a frog and you can see its tongue streak out to snag a fly and its mouth snap shut. Dissect your subject and you can see its throat and esophagus, and how the stomach connects to the intestines. Touch the frog's quadriceps with a pin, and the muscle contracts and the leg moves. The frog's heart, lungs, veins, and arteries are obvious, even if how they coordinate is not immediately clear. But look at a tree, a petunia, or a blade of grass, and you understand nothing of its workings. There is no mouth. If it eats anything, its food is invisible. It produces no waste. Cut it open and you are no more enlightened: There is no stomach, no heart, no digestive system, no musculature. The most you discover is that there are softer and harder tissues inside, and that liquids, clear or milky, may leak from the cut. As for reproduction, it might seem that a single plant produces a seed all by itself. Ferns whose spores are microscopic you might reasonably assume generate spontaneously out of nothing. Certainly, no one could guess by merely looking that plants turn sunlight, air, and water into stems and leaves and flowers.

Uncovering the laws of celestial mechanics, where measurements and equations reveal truths, was easier than understanding an onion. The most fundamental questions about plants were hard to answer. No one ever wondered if a comet was some sort of high-flying bird, but botanists struggled with the basic definition of a plant. Some argued that plants were more like stones than animals. Consider this, they said: Knock a piece off a salt crystal and the chip will grow; clip a stem from a plant and the cutting can develop into a new plant. On the other hand, amputate a man's leg, and it never grows into a new man, and, unlike the crystal and the plant, the man often dies as well. Nonetheless, during the late seventeenth and eighteenth centuries in Europe, in the period known as the Enlightenment, a few men began to look at the vegetal world in a new way, not to see whether plants could be eaten or what diseases they might cure, but to answer a novel question: How do they work? Bit by bit, they revealed the wonders of plant anatomy and physiology.

Those wonders are on display in everyone's garden—and all of Earth is a garden—but some plants are particularly entertaining to partisans of the natural world. A fern that vacuums arsenic out of contaminated soil, a biofuel grass that grows twelve feet tall, utterly black petunias, one-ton pumpkins, and photosynthesizing sea slugs are curiosities, but also instructive about how roots, stems, leaves, and flowers work. Land plants have been evolving for more than 400 million years, more than twice as long as mammals, and they are stunningly diverse in their strategies for survival. Despite their immobility and lack of muscle tissue, they have developed elegant solutions to the problems of gathering food, transporting water and nutrients around their corpus, and reproducing. It's a good thing they have: We depend utterly on plants for food, both by eating them directly and by eating animals that have eaten them. The oxygen we breathe is manufactured by plants and photosynthesizing algae and bacteria.

Plants and their seagoing ancestors have always influenced the global climate, and they will be critical in the coming centuries of climate change. Some of the extraordinary plants I investigated for this book give me hope. *Miscanthus*, that giant grass, is a carbon-neutral biofuel that grows quickly, without fertilizers, noninvasively, and on land too poor for agriculture. The grandparents of plants, photosynthesizing cyanobacteria, can be coaxed via genetic modification into excreting ethanol for fuel. Scientists at the Land Institute in Kansas are developing perennialized versions of annual crops that would sprout anew each spring from underground root systems. By reducing annual tilling of millions of acres, we could slow down topsoil erosion, which is a serious problem in the United States and around the world, and rebuild soil fertility. Researchers at the University of California, Riverside are working to unlock the secrets of a plant hormone that might help crops survive the stress of drought. The International Rice Research Institute is trying to develop rice varieties that incorporate the variant of photosynthesis that makes crabgrass so damnably successful in my backyard.

So, welcome to the garden of marvels, past, present, and future.

A Garden of Marvels

PART I

Inside a Plant

one

Cocktail, Anyone?

It is mid-July in mid-Florida. The sky is one vast sheet of burnished aluminum, pristine except directly overhead where some celestial welder is blazing a large and fiery hole. Fortunately, as I cross the rutted yard from where I've parked my rental car, I feel a light breeze, a minor but meaningful dispensation.

I am here to see Charles Farmer, the owner of this citrus nursery. A small trailer, which I assume is the office since it is the only building in sight, sits on cinder blocks not far from the entrance gate, but no one answers my knock. I worry that perhaps he has forgotten our appointment today, but then I hear a radio playing somewhere out back. Skirting a temporary swamp left from last night's downpour, I follow the sound to a large greenhouse, push through the outer door, and let it close before opening the inner door.

Inside, there is no breeze at all. The air is absolutely still, bright with sunlight and dense with heat concentrated by the translucent roof. Two people are working at the far end of the structure, and I wave and start down toward them, between rows of waist-high benches covered in potted, two-foot-tall saplings. Sweat breaks out on my upper lip. By the time I'm close enough to call hello, my face is slick.

Never mind. I am near the end of a quest, and somewhere in this greenhouse is the object of my desire, a tree that I have been thinking about for years. Somewhere in this structure is a citrus cocktail tree, a single tree whose branches bear many different kinds of citrus fruit.

I had seen such a marvel growing in the dirt floor of an old greenhouse at Logee's Greenhouses in Danielson, Connecticut, when I was interviewing the owner for my previous book. It was magnificent, its leafy limbs brushing the glass roof and stippled with oranges, tangerines, lemons, limes, grapefruits, and kumquats, as well as hybrids like sunquats and limequats. It seemed the kind of tree that would harbor a grinning Cheshire cat or shade a blue caterpillar smoking a hookah. Possibly, it was magical: Eat one of the fruits and you live happily ever after.

I wanted one.

I wanted one not only because I have a weakness for the charmingly improbable, the scientifically curious, but because for me it would have a practical value. I have a collection of potted citrus that lives in our small conservatory in the winter. The conservatory faces north, not an ideal orientation, so I have to hang fluorescent grow lights to keep them alive through the season. For many years, I could gather all my trees under five long grow-lights where, lush and green and still holding the fruit of summer, they help me get through the dark days of winter. I'm a sucker, though, for trying unusual citrus varieties, so now three of the largest trees have to overwinter in the garage. There they go dormant and drop their leaves, looking like death until I rusticate them again in March, when light and warmth resurrect them. With a citrus cocktail tree, I could keep more varieties flowering and fruiting indoors when I'm most in need of their tonic beauty.

I set about trying to buy one, and quickly discovered that no nurseries or garden centers in Maryland or Virginia carry a cocktail tree. No one had even heard of such a thing, although everyone found the idea intriguing. Logee's had none, either, so I went farther afield and called retail nurseries in Florida, hoping someone might ship me one. Although employees there had heard of the tree, no one had one to sell. Finally, in January, I put in a call to the University of Florida Citrus Research and Education Center, and a professor suggested that I call Charles Farmer, owner of a citrus nursery in Auburndale, a town halfway between Tampa and Orlando. If anyone had a cocktail tree, he said, it would be Charles.

But not even Charles had one.

"Used to be some call for 'em," he said over the phone in a drawl that sounded like Southern Comfort cut with a generous squeeze of lime. "Home Depot ordered a thousand back in '92, but then canceled on me. It's no problem, though, for me to make you one."

It's no problem, I discovered, because Charles is in the business of making citrus trees, about a hundred thousand

of them every year. I use "making" advisedly. Most commercial citrus trees are not grown from seeds. Plant a seed
you've found in a tangerine, for example, and if the little sapling doesn't first succumb to bacteria, viruses, nematodes, or
fungi, you will have to wait years for it to grow large enough
to fruit. Even when your tree reaches maturity, its tangerines
will probably be sparse on the boughs and smaller than what
you can find at the grocery.

Instead of planting seeds, commercial growers buy grafted
saplings, either to start a new grove or to fill in an existing
one where individuals have died. Grafted trees like Charles's
are made of a hardy rootstock species—often a sour orange—
onto which is grafted a more commercially desirable variety
(the "scion"), like a Satsuma or a blood orange. The rootstock
imparts vigor, disease resistance, or cold tolerance, or a combination of those qualities to the fruit-bearing branches. The
rootstock can also improve the yield of the scion or intensify
the flavor of its fruit. Of course, seedless citrus varieties must
be grafted to reproduce at all.

Charles assured me he could easily ship me a cocktail
tree. I was thrilled. Could he ship it in its pot, or would that
be too heavy? I hoped he wouldn't have to bare-root it, I told
him, because I was afraid I would kill it in the repotting.

He laughed. Shipping was no problem because what
he intended to send me was a "liner." A liner is a twenty-
four-inch-high grafted sapling with a trunk like a pencil and
rooted in a four-inch-diameter, eighteen-inch-deep pot.

I tried not to sound disappointed since this seemed to be
my best chance for a cocktail tree. But such a little specimen,
especially at my northerly latitude and indoors during the

winter, would take years to produce any fruit. Was there any way I could get a bigger tree?

"Sure thing," he said, "if you're willing to come down here and cart it back home. I got a nice six-foot Hamlin orange in a thirty-five-gallon pot that I can do for you. It'd be ready next spring. That fast enough for you?"

Actually, it was not. That was fifteen months away. I imagined Charles attaching slender branches to the thick trunk of the Hamlin the way a surgeon might reattach amputated fingers onto a hand. I pictured a heavy wrapping of bandages around the juncture of old and new wood, and fleetingly wondered whether he used wires to hold a branch in place while the new joint healed. Surely, the tissues would knit together and wounds would close in a matter of weeks or, at the most, a couple of months, as they did in skin. It shouldn't, it seemed to me, take more than a year for a connection to form. Why was it, I asked tentatively, that I couldn't get it sooner?

Charles began to explain, but I didn't understand what he was saying. Then, suddenly I got it: My analogy was wrong. He would be grafting buds, not branches, onto the Hamlin. Grafted buds "take" in a couple of weeks, but they need a year or more in Florida's climate to develop from bud to branch. And the best season for grafting is in spring, when the trees are actively growing and sap is flowing. Hence the fifteen months.

Charles added that the bulk of his business is producing liners in his greenhouses. From time to time, however, he and Susan, his wife and business partner, get a call to "topwork" mature trees in a grove, transforming them from one variety to another. Recently, for example, when the market price of navel oranges fell and seemed unlikely to recover, a farmer nearby hired him to turn fifteen acres of navel oranges into fifteen acres of Valencias. Topworking is expensive, as might be expected for such major magic, but, he said, "it sure beats bulldozing and starting from scratch." Topworking was what he planned to do to a Hamlin orange tree for me. The only question he had was what varieties I wanted on my tree. I wasn't particular, I said, as long as I got a range of fruit colors. I didn't plan on eating my crop—I can buy fruit in the grocery store—I just wanted to look at it. He chuckled, and said okay. I have no doubt that no one had ever asked him to graft a tree for its aesthetic appeal.

I announced (well, maybe I exulted) to my friend Edie, a buyer at a local garden center, that I'd found someone who would make me a cocktail tree. She looked skeptical. I'd better be careful, she said; there were restrictions on taking citrus trees out of Florida. That seemed hard to credit—wasn't Florida all about shipping citrus?—but I looked into it and found that indeed the U.S. Department of Agriculture prohibits the transportation of citrus trees out of the state. The regulation is an effort to prevent a bacterial disease called "the greening" from spreading beyond Florida to other citrus-producing states. The greening, also known as Huanglongbing or the yellow dragon disease, is caused by a bacterial infection and transmitted by a plant louse. The infection causes leaves to turn yellow and fall off, sours the fruit, and ultimately kills the host. Violations of the regulation are punishable by fines from $500 to $10,000. There was a bit of hope, though. Shortly before I made this disheartening discovery, the USDA proposed regulations that would allow licensed growers, by virtue of taking extra precautions and undergoing additional inspections, to earn a certificate to ship out of state.

When I called Charles to tell him about the problem, I got an earful. He was already a licensed grower, he already followed all the costly procedures to ensure his trees were free of disease, he already was inspected regularly. He had no interest in more paperwork, more fees, and more inspectors poking around. The whole thing was ridiculous since I would be driving directly to Maryland, where there is no citrus industry and not even any backyard citrus trees. The regulations, he insisted, made no sense in my case. I told him I'd contact the USDA and see if they ever issued waivers. I called, but the people at the Animal and Plant Health Inspection Service (APHIS) were clear: no exceptions, not even for a Marylander who promised to drive with her windows rolled up and not to stop—not even to eat, much less stay overnight—in Georgia or South Carolina where citrus are cultivated or even North Carolina or Virginia where they are not.

When I phoned Charles a third time, he really growled. Back in the mid-1980s, he recalled, the regulations on budding were tough due to a canker outbreak. "They actually had *Yankees* come down that didn't know anything about citrus that was watching us professionals. We had to get a 'budding card' and show it every time some inspector come by. And inspectors'd park out on the road, and we'd have to walk all the way back out from where we was budding, over that burning hot sugar sand, every time they showed up."

I thought I was out of luck, that he'd never agree to apply for the additional certificate, but I misunderstood his temperament. Having sent a little steam through the ether, he then amiably agreed to help me out, me, a Yankee he'd never even met. He would make the application whenever the regulations became final. And he and Susan would take care of my tree in the meantime. Since I fly regularly to Florida to visit my mother, who lives about two hours south, in Fort Myers, I asked if I could drive up and watch him graft my tree the next time I was in the state.

And so it is that I am in this greenhouse sweating like a sprinkler system and shaking hands with Charles and Susan. Both look to be about fifty, and are lean and deeply tanned. Charles has a square jaw and prominent cheekbones (thanks, I learn, to his Osage Indian heritage), emphasized by his shaved head. He is dressed in baggy, knee-length shorts and a faded Hawaiian shirt and has a wide smile; Susan wears shorts and a white halter top, and a visor wraps around her sun-streaked hair. Both hold what look like paring knives and have bumpy rubber sheaths on their thumbs. Susan has several ragged bandages around her fingers.

After we introduce ourselves, Susan announces that she's going to get the budsticks from the refrigerator in the office, and I follow Charles, who is heading to the far end of the greenhouse. I hustle, dripping, to keep up. Our destination is a row of lush and leafy six- and seven-foot trees in black plastic tubs.

"These here are our budwood trees for Hamlin orange," Charles says. "They're certified disease-free by the state. When we get an order for Hamlins, we take buds from these trees and graft them onto a Swingle rootstock." Swingle is a particularly vigorous variety of citrumelo, which is itself a cross between an inedible grapefruit and an inedible orange. If a Swingle were to grow to maturity, its fruit would be small, sour, and full of seeds.

Charles heads to a tree in the row. It looks as if a hurricane has blown through this particular individual and stripped it of most of its small branches and leaves. It's a skeleton clothed in rags. This, it turns out, is my tree.

I'm dismayed, but Charles tells me with obvious satisfaction that he chose it specifically for me because it has five good "scaffold" limbs. (Scaffold limbs are like railroad trunk lines off which all the local lines run.) My tree hasn't lost its foliage; Charles has stripped it down in preparation for grafting. On each of the scaffold limbs, he will graft two buds from five different citrus species. The remaining Hamlin branches and leaves will sustain the tree while the grafts grow out. Then he'll remove what is left of the Hamlin beyond the trunk.

Susan arrives with the budsticks, which are also certified disease-free, in individual plastic bags. She and Charles have chosen Meyer lemon, Eureka variegated lemon, Bearss lime, Cara Cara navel, and Minneola tangelo. The budsticks are bumpy foot-long pieces of quarter-inch-diameter stem stripped of leaves and twigs.

Charles explains. "Look in here," he says, pulling down a branch of the Hamlin and pointing into the angle—the armpit, you might say—between a branch and a leaf stem.

"See that?" he asks. I see a pointed bump. "That's an axillary leaf bud, or what we call the eye. The tree doesn't need the eye to grow now so it's dormant. It has the potential, though, to become a new branch. The budsticks also have eyes and what we're going to do is take eyes off 'em, and use a chip graft to put 'em into the branches of the Hamlin."

Charles takes his knife to the Hamlin, first making an inch-long vertical cut through the bark into the wood, then slicing horizontally to create an upside-down T. Then he turns to the budstick, and shows me the eye he plans to excise. Holding the budstick so its upper end points away from him, he draws the knife toward his body to make a shallow inch-long slice—the chip—just under the bark surrounding a bud. Finally, he slips the bud (pointing upward) beneath the bark flaps of the Hamlin, takes a roll of green plastic tape from his pocket, and wraps the wound with a length of tape.

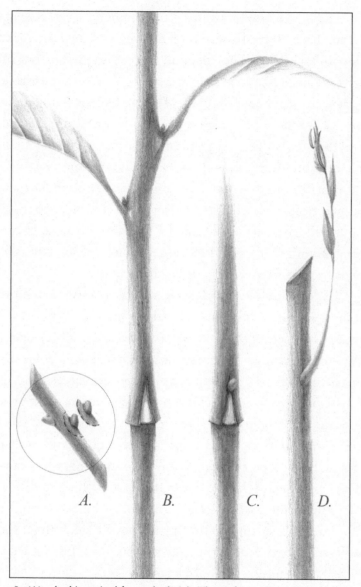

In (A), a bud is excised from a budstick. The grafter makes an inverted T
in the bark of the rootstock (B), and slips the bud under the flaps (C),
before wrapping the graft with tape. In (D), the cambium layers have
fused and the bud is fully integrated.

The key, he says, is getting the cambium layers, which lie just beneath the bark in both the bud and the rootstock, close to each other. The cambium is exceedingly thin, but its cells are much like stem cells in animals, that is, they are able to generate different cell types. (The cambium is too shallow to observe without a microscope, but in the spring, if you peel back the bark on a twig, the moist surface is cambium.) In a successful graft, cambium cells link tissues and vessels in the rootstock to the equivalent tissues and vessels in the bud, permanently connecting them. In two weeks, those connections will have been made, and he'll remove the tape. In the bright sunlight of central Florida, it won't be long until the buds become slender, leafy young branches.

Ten minutes later (and going slowly so I can watch his technique), Charles is finished making my tree. With a lifetime of practice, he can make a graft in seconds. What's more, 99 percent of the grafts that he and Susan make—and he estimates they've made more than a million—are successful.

Before I go, we sit down at a picnic table sheltered by a majestic live oak, the only shade tree on the property, to talk, and Susan offers to get us some "ostee." I don't know what that is, but accept, thinking it might be something like candied citrus peel. What I'd really like is something to drink, so I'm delighted when *ostee* turns out to be Southern for *iced tea,* sweet enough to fur the tongue. We laugh about the differences between our accents and cultures, but actually, Charles tells me, he was born a northerner.

"Mom and Dad was on their way back from Michigan where they was picking apples. They was migratory, so they would start on cotton in the south, then tobacco, cherries, and apples, doing the full route south to north. Well, they ended up on the apples, and was on their way back to Florida to start buddin' when Mom went into labor in Cincinnati. After three or four days, they turned her loose, and we shot straight on down here. Been here," he said with satisfaction, "ever since." Well, he amends, there was one year of college at Kent State in Ohio before he saw the error of his ways.

Charles and Susan have been in the citrus business for more than thirty years now, first as budders in others' groves, and then on their own property and in their own greenhouses. It's been no easy life. The work is hot and tedious; outdoors the mosquitoes and gnats can be fierce; and they are constantly pierced by thorns. Few Americans are willing to help them do it, and even immigrant labor is tough to come by. Citrus diseases and the resulting quarantines have put them out of action more than once. Freezes have hurt, too. For years, Charles had to go to the Bahamas to bud in order to keep them afloat financially.

"This past winter," Susan says, "our lemons and limes was just poleaxed, down to the ground. The problem was it had been in the nineties when the freeze hit. The trees wasn't dormant and when the cold hit, the sap was flowing in them." (Frost damages plants when the liquid inside individual cells freezes and forms ice crystals that rupture the cell walls. When later the ice melts, the fluids drain out, so death actually comes from dehydration.) They had to bulldoze four acres of trees. Recently, they've added blueberry bushes to the property, in an attempt at diversification.

Later, as the two of them walk me to my car, I ask Susan how she knows when it's time to water her trees. She tells me it's easy: She gives each pot a kick every day, and she can hear by the sound. Before I can formulate a follow-up question—what exactly are the sounds of dryness—she reminds me I'll have to prune the lemon branches hard. Charles had grafted the lemon buds onto the narrowest limbs to give the other varieties a head start, but lemon species are vigorous and will quickly shade out the other varieties if I'm not careful. Then Charles unlocks the chain-link gate, and I drive carefully between the ruts in the dirt yard. In no time at all, I'm heading out along Interstate 4 back to Fort Myers, air-conditioning on high, as visions of citrus fruits—yellow, pink, orange, red, and striped—dance in my head.

two

The Birth and Long Life of the Vegetable Lamb

My fruit cocktail tree is man-made, but theoretically one could develop naturally. Wild grafts can arise when limbs from two genetically similar trees—a lemon and a lime, for example, or two apple varieties, or two stone fruits, like a nectarine and a peach—come in continuous contact with each other. As the limbs rub together in the wind, the bark of both wears away at the point of contact, and ultimately their exposed cambium layers fuse. Cut either branch free of its parent and it will rely on the root system of its neighbor to prosper. Since the medieval era, gardeners have used this natural phenomenon to create living fences. "Pleaching" involves interweaving the branches of a line of young trees or shrubs, scoring through the barks to the cambiums where they intersect, and then tying the intersections in place to ensure they graft.

A fruit cocktail tree, therefore, is improbable but not magical. The people I asked at the Maryland garden centers who had never heard of such a tree intuited this. The idea intrigued and amused them, but the tree did not seem outside the realm of possibility. It wasn't as if I had asked them for a tree that produces kittens, bear cubs, or little lambs, a possibility no one would entertain for a second. Today, no matter how slight your formal botanic education is, you have no doubt that plants cannot sprout baby sheep.

Not so in the early seventeenth century. Educated Europeans believed that out on the central Asian steppe there were plants that grew tiny, perfect, living lambs. This "vegetable-lamb," also known as a "borametz" or "barometz" (Scythian for sheep), was said to emerge from the top of the plant's central stalk, to which it was attached by its navel. The lamb's four cloven hooves hung down, but not far enough to touch the ground. Fortunately for the borametz, its stalk was flexible and its neck was long enough so that it could, by leaning this way and that, nibble the grass in a circle around the base of the stalk. Unfortunately, once it had eaten all the grass within range, the stalk withered and the little lamb expired. Once the lamb died—either from starvation or at the hands of hunters—its soft and surpassingly white fleece could be sheared, and woven into the fine, snowy cloth for which the region was famous.

The mythical borametz.

Much detail was reported about the borametz. Sigismund, Baron von Herberstein, the ambassador to Russia under Holy Roman Emperors Maximilian I and Charles V, wrote in 1549 that he heard from unimpeachable sources that the creature "was of so excellent a flavor that it was the favorite food of wolves and other rapacious animals." Johann Bauhin, a respected Swiss naturalist, wrote in his *Historia plantarum universalis* (circa 1600) that its blood is "sweet as honey, and its taste is like the flesh of fish." Others opined that it tasted more like crayfish and that only wolves, not other carnivores, would attack it. Claude Duret, author of the *Histoire Admirable des Plantes* (1605), classified the borametz as one of several known "plant-animals," a group that also included sea nettles, sea sponges, and sea lungs. John Parkinson, apothecary to King James I and the greatest English gardener of his era, published a comprehensive treatise on horticulture in 1629. The frontispiece of his book was an engraving of the Garden of Eden, a locale universally accepted as real and perhaps still extant. In Parkinson's Garden are familiar species like lilies and asters. Others are Old World imports like date palm, cactus, and citrus. At least one, the pineapple, is from the Americas. And there, center-left, is a borametz, its head and legs dangling, atop its stalk.

To believe in a lamb-producing plant—or barnacles that grow on trees and then hatch geese, asparagus that spring from crushed rams' horns, or mandrake roots that scream like men when pulled from the earth—seems laughable now. But in the seventeenth century, no one knew what a plant was or was not. An intuitive definition of plants—something along the lines of "living, green, immobile, insensate beings"—was of little help in delimiting the category. It was easy to find ferns and trees with brown or reddish leaves. The dodder, orange and leafless (and which lives as a parasite on other plants), was prescribed by physicians and healers to cure melancholy. Immobility was a slippery concept: Petals open and close, leaves turn toward the sun, and tendrils climb and twist with almost visible speed. Not all plants are insensate: Touch the sensitive plant (*Mimosa pudica)* and its pairs of little leaves immediately fold in sequence.

And what about those odd sea creatures that Duret noted? They seemed to straddle the worlds of plant and animal. Visible in shallow waters, they appear to be rooted to rocks or coral, and many have branchlike or leafy appendages that sway in sea currents just as terrestrial plants' branches bend in the wind. They are often earthy hues. On the other hand, some might respond to a touch, and others were observed to consume tiny shrimp and even small fish, a decidedly animal-like trait. It was not illogical to conclude that these sea creatures were a combination of plant and animal. A borametz was simply a terrestrial version.

In fact, it wasn't clear what the dividing line was between living and non-living entities, much less between a plant and an animal. A depleted mine might yield new ore when reopened years later, which seemed proof that stones could grow. According to Aristotle, the great Thales of Miletus attributed the attractive power of magnets to their souls, and souls, everyone would have agreed, could be found only in living things. Medieval alchemy was predicated on the belief that base metals slowly grow in the earth, maturing by stages from lead to tin to copper to silver until they reach adulthood as gold. This seemed like a reasonable theory: After all, inanimate eggs turn into chicks and inanimate seeds transform into miniature plants. Alchemists designed their elaborate procedures in accord with abstruse theories coded with astrological symbols, but they always used heat in their procedures, burying lead or tin in warm manure, roasting them in ovens, or simmering them in flasks. Heat and growth seemed to go together. Eggs required the warmth of a hen's body to develop and seeds needed the sun's warmth to sprout. Likewise, metals sluggish from living in the cold earth had to be warmed to ripen.

It was easy to see salt and sugar crystals grow, which seemed to some an indication of a type of life. The dendritic formations on stone that look like fossil plants (and are actually crystalline structures) were a source of great confusion: Were they plants growing out of stone or were they plants made of stone? Miners found twisted veins of copper and silver that looked like plant stalks: Did metals grow like plants or were these metallic plants? Certain mosses, it was noted, could go dry for months and even years—long enough that anyone would assume they were dead—but when they were wetted, they became green again. Had they died and been resurrected? Or perhaps they had never lived in the same way that plants and animals live? Farmers observed that plants were changeable entities: Rye seeds planted in a field occasionally emerged as barley or, worse, useless weeds, especially after heavy rains. The boundaries between plant and plant, plant and animal, and even between inanimate and animate were uncertain. Just as it was in a heavenly rainbow where red bleeds into orange and indigo becomes by imperceptible degrees violet, so it was on Earth. A rhubarb was a plant, a sheep was an animal, and a borametz was somewhere along the continuum between the two. Like a fruit cocktail tree in the twenty-first century, the particular characteristics of the borametz were curious, but reports of its existence were unsurprising.*

In 1698, Sir Hans Sloane, secretary of the Royal Society in London, slew the borametz. The Society had recently received a collection of natural curiosities and medical implements from China. Sloane was particularly fascinated by one object:

> *More than a foot long, as big as one's wrist . . .*
> *covered with a down of a dark yellowish snuff color,*
> *some of it a quarter inch thick . . . It seemed to be*
> *shaped by art to imitate a lamb, the roots or climbing*
> *parts being made to resemble the body, and the*
> *extant foot-stalks, the legs.*

The object, he concluded in an article published in the Society's *Philosophical Transactions*, was a piece of the root of a tree fern, and was also the source of the story of the borametz.

Sloane was correct that the lamblike object was formed from the root of a tree fern. He had seen similar ferns in Jamaica during his stay there as personal physician to the governor, the Duke of Albemarle, in 1688. He was wrong, however, about the origin of the borametz myth. The tree fern is tropical, and could not possibly grow on the cool,

---

* The line between animate and inanimate is still is not perfectly clear, and modern science recognizes unusual organisms that straddle two categories. Outside a host cell, a virus is known as a virion and is simply an organic (that is, carbon-based) particle with no internal biologic activities. When a virion comes into contact with a host, it becomes a living virus, reacting to its environment and attempting to replicate. Certain algae are known as mixotrophs. As plants, they produce their own food through photosynthesis, but sometimes they sustain themselves by engulfing and digesting other plants and even their former predators, and then are classified as animals.

dry steppe where the vegetable lamb was said to graze. (The fern now immortalizes Sloane's error: Its botanical name is *Dicksonia*—for James Dickson, a Scottish botanist— *barometz*.) The pedigree of the borametz, however, actually lies in ancient Greece, in the writings of Theophrastus, the philosopher who taught at the Lyceum, the school of philosophy and science founded by Aristotle in Athens.

Theophrastus was born on the island of Lesbos in 371 B.C., but as a young man, he moved to Athens to study, probably with Plato. He met Aristotle, fifteen years his elder, who tutored him in philosophy and natural history. Aristotle is said to have given his gifted and persuasive pupil his name: *Theo* means divine and *phrastos* means expression. The men became intellectual collaborators at the Lyceum, and Aristotle would appoint his former pupil his successor as the school's director, as well as the guardian of his children. About 343 B.C., the two scholars undertook a comprehensive assessment of the natural world. Aristotle focused on animals while Theophrastus devoted himself to plants.

In the two volumes of *De Plantis*, Theophrastus treats questions of plant disease, medicinal uses of plants, methods of sowing and pruning, and other topics of practical concern to students in an era when wealth was based on farm holdings. As for grafting, a procedure familiar to growers of figs and olives in ancient Greece, he reported that it was best to join "like on like," and observed that the greater the similarity between two trees' barks, fruits, habitats, and growth cycles, the more likely the graft would be successful. He essayed an early taxonomy, grouping plants into categories to try to make sense of their diversity. But, most of all, he observed—carefully, minutely, and dispassionately—trying to determine what plants truly are.

His book starts with this declaration: "We must consider the distinctive characters and the general nature of plants from the point of view of their form and structure, their behavior in the environment, how they reproduce, and the general course of their life." His topics couldn't have been more modern; in fact, they correspond to the chapter headings of a botany textbook I have. At a time when plants were of interest only as medicine and food, when pieces of wood that sprouted shoots were portents, and when a laurel tree or a hyacinth might be a winsome girl or handsome young man transformed by a god, Theophrastus's approach was remarkable.

He was the first to describe plants analytically, describing and differentiating root, stem, branch, twig, leaf, leaf stem *(petiole)*, bark, wood, pith, sap, and fruit parts, among others. He noted that leaves are attached to stems in distinct patterns, that buds arise in order and in particular places, and that new growth can arise from the top of the plant or from the side. Emerging seedlings have either one primordial leaf or two, algae have no roots, and ferns have no seeds. He described various trees, and could say whether their bark is easy or difficult to strip, if their wood rots quickly or not, and when exactly they fruit and drop their leaves. He measured the depth of roots, the length of branches, and the height of trunks, and noted how wild forms differ from their cultivated counterparts, whether plants flourish in marsh or meadow, and how temperature affects their growth, among dozens of other bits of ecological information. Date palm growers, he reported, sprinkled the "dust" of flowers that grow on male trees over the flowers on the female trees.

If no one performed this dusting, fewer dates would develop. Above all, he saw plants as intrinsically interesting.

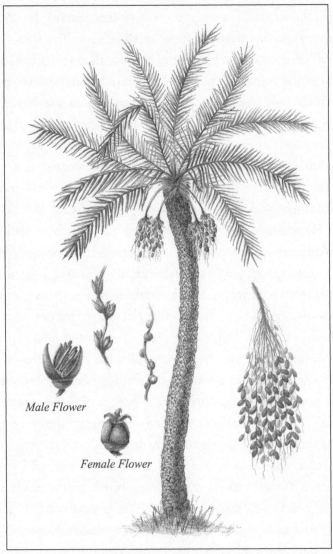

North African date palm.

Theophrastus rejected ideas based on evidence that is "beyond the ken of our senses." Pieces of wood that sprouted shoots were not signs or "anything unreasonable." The bay laurel found growing in the crotch of a plane tree was not evidence of a divine transformation: A bird must have dropped a seed in some moldering leaves. His goal was to rely on the "admitted and observable" for understanding. He rejected the theories of Anaxagoras, who believed that air contained seeds that fell to the ground when it rained, and Diogenes, who thought plants were produced when water decomposed and mixed with earth. While he didn't question that seeds might change species in the ground, what he did affirmatively accept—floods carrying seeds to new areas—was a process he could see.

Theophrastus wrote about certain small, "wool-bearing" trees that grew in India and Arabia that "bear no fruit, but the pod containing the wool is about the size of a spring apple. . . . When it is ripe, it opens, and the wool is then gathered from it and made into cloth." The Greek word he used for apple (μᾶλον, pronounced "mal-on" and the ancestor of our "melon") had three meanings: apple, fruit in general, and sheep. At some point, "spring apple"—that is, a young and small apple—was understood to mean a "spring sheep," that is, a lamb. Since Europeans had never seen cotton plants, and their only experience of a fluffy, white substance that could be woven into superfine cloth was lambs' wool, the vegetable-lamb was born. It is one of history's ironies that the most science-minded philosopher of the ancient world, a man who treasured accurate observation of the plant kingdom, was responsible for the myth of the borametz.

If Aristotle had been the one to write the book on plants, he might have started with "First, we must consider the purpose of a plant." Of barley, he might have written that the Prime Mover created it with the purpose of producing an edible grain, which is why it has a seedhead that ripens and can be easily harvested. (The doctrine that all things are as they are because a divine being made them for an express purpose is known as teleology.) On the occasions when he did write about the vegetal world, he analogized between plant and animal parts: Roots were like mouths, branches like arms and legs, and leaves like hair. He was not alone in thinking plants were a kind of imperfect animal, made of the same elements only with impurities and without animal warmth.

To Theophrastus, trying to understand plants by comparing them to animals had no value. "It is a waste of time," he wrote, "to take great pains to make comparisons where that is impossible, and in so doing we lose sight of the proper subject of enquiry." Yes, if one looked at the overall form and structure (that is, the morphology) of a plant, there are similarities between the two, but look at growth habits and the resemblance ends. He realized that animals and their parts— the head, legs, arms, etc.—grow to a predetermined size and then stop. On the other hand, while the stems and trunk of a plant may cease growing when they reach a particular size, the plant keeps growing. (Plants have what botanists call an *indeterminate* growth type, meaning that although each one grows according to a genetically determined pattern, new roots and shoots develop as long as the plant is alive.) A plant, he wrote, is unlike an animal and can be pruned, divided in half to become two plants, or propagated from a piece of itself, say a stem or a leaf. We may call plant parts by the names of animal anatomy for convenience, he said, but do not be misled: They are fundamentally different than animals.

If only *De Plantis* had survived. But Latin replaced Greek as the lingua franca of literate Europe in the first centuries A.D., and many of the Greek works, written on delicate papyrus, disintegrated. Theophrastus's *De Plantis,* like the vast bulk of ancient Greek works, disappeared. Instead, for roughly fifteen hundred years, Europeans turned to two Roman authors on plants: Nicolaus of Damascus, a historian at the court of Herod, and the historian and encyclopedist, Pliny the Elder. If you had set out to hold back the progress of botany, you couldn't have done better than Nicolaus. Nicolaus pieced together a hodgepodge of misinformation (also called *De Plantis)* in about 30 B.C. that was misattributed to Aristotle and thereby acquired great authority. Pliny's work, *Natural History,* is a more complicated matter.

Pliny was one of the hardest-working, most compulsive, and most prolific writers in history. Born in A.D. 23, he spent the first part of his career in the Roman army as a staff officer and then commander in Germany. On the German front, he found time at night and during the winter breaks from campaigning to write a history of the Roman-Germanic wars, no small undertaking. In 54, Nero came to power, and the imperial secret police began pursuing the emperor's many real and imagined enemies. Government officials, and especially military officers, suspected of opposition were murdered in the streets, put on trial, or forced to commit suicide. On his return to Rome, Pliny concluded this was the moment to retire from public life, and devoted himself to writing books on Latin grammar, a subject not even Nero could find offensive.

After Nero's death in 68, Pliny, thanks to his early support of the new emperor, Vespasian, and a friendship with his oldest son, was appointed as a procurator in several of the North African and Western European provinces. In this second round of public service, disciplined as ever, he wrote a thirty-one-volume history of Rome, and then the thirty-seven volumes of *Natural History*. For this, his magnum opus and what some call the world's first encyclopedia, he read two thousand Greek and Roman texts and consulted with untold numbers of farmers, craftsmen, tradespeople, and other authorities he came across in his foreign travels. According to his nephew and biographer, Pliny the Younger, he rarely slept and had a lector who read to him at all hours, including during meals, as well as a secretary who was constantly at his side to take notes. In Rome, porters carried him about in a sedan chair so he could read and write and not waste a minute.

Pliny was a magpie of a researcher, and *Natural History* is a catalog of facts about the universe from astronomy to zoology, from winemaking to woodcutting, from magic potions to metallurgy. What constituted a fact—and Pliny claimed his work contained twenty thousand of them—was a matter of opinion. Open his multivolume work to any page and you encounter a heterogeneous accretion of truths and half-truths, observations and hearsay, and myths and mistakes considered at that time to be the realities of the physical world.

As for plants, you need go no further than his first pages on trees to see how it is with Pliny. In rapid order, he gives the reader a history of tree worship, a discourse on the nature of plane trees, the dimensions of four renowned plane trees, a tale of the obese Caligula holding a banquet for fifteen guests and their servants inside the hollow trunk of a plane tree, and a report that plane trees grow faster when watered with wine. He accurately describes the Indian banyan tree with its aerial roots that grow into new trunks but then adds that its "broad leaves have the shape of an Amazon's shield and cover the fruit, so hindering its growth." When it comes to grafting, Pliny records the techniques of his time, techniques not so different from those that Charles and Susan Farmer employ. But he also writes that he has seen a tree "laden with fruit of every kind, nuts on one branch, berries on another, while in other places hung grapes, pears, figs, pomegranates and various sorts of apples." While there are certainly some surprising affinities in the plant world—tomatoes, for example, can be grafted onto potatoes because both are members of the Solanaceae family—no rootstock is compatible with all these unrelated species. This cocktail tree, he notes, didn't live long, but he didn't suspect, as was no doubt the case, that some trickster had cobbled it together with a sharp knife and a cup of resin.

It wasn't only the errors in *Natural History* or the pseudo-Aristotle's *De Plantis* that held back progress in botany, but how deeply the scholars of the medieval era revered these works. Anyone with an interest in understanding plants simply turned to the ancients who, it was accepted, had already provided the definitive information. No one had yet conceived the idea of experimentation as a method for discovering facts or for testing the validity of a hypothetical truth.

Alchemists, physicians, and farmers experimented, but only in the sense of trying alternative means to achieve an end. In addition, medieval scholars were imbued with Plato's belief that in the divine mind there exists an ideal form, the "universal form," of every creature and thing. The lamb or the orange tree that we see on Earth is only a shadow, a simulacrum of the ideal animal and ideal tree. If a student wanted to understand an orange tree, he would not be advised to examine a real—and therefore defective—specimen, but instead to reason his way to knowledge. This philosophy, as embraced and interpreted by the medieval Catholic Church and then the late-medieval universities, bred a mind-set disinclined to making close observations. As a result, an understanding of how plants functioned progressed not at all from the third century B.C. to the seventeenth century, and vegetable lambs continued to bloom and bob in the landscape.*

---

* Taxonomy, or the science of classification, of plants is a different matter. In the Renaissance, especially as European explorers brought home new species from Asia, Africa, and the Americas, people became absorbed in collecting, describing, displaying, naming, drying, painting, and categorizing plants. Carol Kaesuk Yoon explores the general history of taxonomy in *Naming Nature: The Clash between Instinct and Science* (New York: Norton, 2009); Anna Pavord recounts the early history of plant taxonomy in *The Naming of Names: The Search for Order in the World of Plants* (New York: Bloomsbury, 2005).

three

Through a Glass,
However Darkly

If there was a science course in the seventh grade at Pimlico Junior High School in Baltimore, I have no memory of it. Of course, few of my memories of that year had to do with anything academic. I was a quiet, slight, prepubescent twelve-year-old fresh from a small elementary school in a white neighborhood plunked into a redbrick factory of a building, designed for 1,500 but jammed with 3,000, most of whom were from poor, black neighborhoods. I was too callow to appreciate it at the time, but the explosive racial anger that would burn a good portion of the city the next year, in April 1968, smoldered within the school walls. I did understand that school was effectively segregated, with the minority white students separated in nearly all-white classes. It seems too grotesque to be true, but my recollection is that white students were in the "A" track while the majority of black students were in either the "B" or the "Basic" track.

Lunch in the cafeteria was a twenty-minute foray into a general tumult that from time to time would erupt in a melee. I tried not to drink anything at lunch, the better to avoid a trip later to the bathroom where older girls cut classes, jived, and smoked. I dreaded the bell at the end of a period because it meant venturing again into the torrent of kids that rushed and roared down the hallways. Teachers, too, were on the move. They didn't have their own classrooms but rolled their books and materials on a cart onto the elevator and from room to room. Adults were posted at major hallway intersections to direct the flow, temporarily halting one seething river to allow another to pass. Who knew what turbulence would otherwise result at the junction? The windowless staircases, however, were unmonitored, and there the human traffic would back up. Jammed together, a girl (and this was one of the few race-neutral phenomena at the school) was subject to being stuck from behind by a boy wielding a sprung safety pin. When it happened to me, I would twist around to try to confront my tormentor, but no gleeful or malign expression gave away the single culprit among the many bodies around me. The only thing to do was to wriggle to try to make oneself a more elusive target until the logjam broke. The best strategy, I learned, was to hug the wall when going up the stairs.

At the end of the school year, half the students were sent by class, younger grades first, to the auditorium for an assembly. We gathered in ever-growing numbers in the spacious foyer outside the multiple sets of doors, doors that someone had yet to, or forgotten to, unlock. At first, we talked and laughed—it was the end of the day and nearly the beginning of summer—but the foyer soon filled, and the air grew warm and fuggy and voices grew louder and impatient. Then a group of bigger, older boys, arriving late, began for sport to push rhythmically against the crowd in front of them, which was hemmed in by three walls. Everyone shuffled forward and then struggled to step back, as we were compressed again and again. There were shouts of protest; girls screamed. With my face pressed against the back of the boy in front of me, I could see nothing but plaid. The corner of someone's notebook gouged my shoulder. At one surge, my feet lost contact with the floor. I dropped my books to grab my neighbors' arms; I was terrified I'd slip down and be trampled. I don't recall how this drama, this mini-riot, ended, but I know I lost my shoes as well as my books. My parents, firm believers in public education, lost their nerve, and managed, with tuition help from my grandparents, to get me into an independent school for the next year.

I found my academic feet at the Park School, a progressive institution on eighty acres in Baltimore County where the chief education goals were "development of the individual" and support for "outlets for creative expression." At the time, there were about fifty students per grade, but this was enough to produce a full-length opera (written by the music teacher) each year and several plays. An art teacher directed a crew of student set painters for these productions in a barn on the campus. A newspaper, an alternative newspaper, and a literary magazine flourished, as did drama and film clubs. I fell in love with American and English literature, especially poetry, found a mentor, and became convinced I was destined to be a modern Wordsworth. (One of my efforts of this era was titled "Imitations of Immorality," based on my naïve belief that the stories told by certain classmates about their weekend escapades were mere braggadocio.) I read and wrote poetry in all my spare moments, as well as moments that my parents and teachers did not consider mine to spare.

In this period of my self-directed apprenticeship, I came to believe that learning science and mathematics was antithetical, and possibly detrimental, to my literary aims. Also, it was hard work, and I didn't have the natural facility for it that some of my classmates did. Rather than struggle for mediocrity, I pursued the humanities and avoided math and science as best I could. I took five years of French and four of Russian, five years of history, every literature course offered, and then tutorials when I'd taken all the courses. I managed to get through the requisite year of geometry and two years of algebra by sitting in the front row and nodding sagely, as if I found the proof chalked on the blackboard entirely self-evident. Luckily for me, at Park comprehension of mathematical concepts (the only evidence of which was my nodding) was deemed as important as actually getting the correct answers on tests. Only two years of upper-school science were required, so after taking Earth science in eighth grade, I took biology and chemistry, and quit the field before my measure was truly taken.

My difficulties with science were more than conceptual. The science teachers, who were not Platonists, assumed that in order to learn the subject matter students had to do more than read and listen; they had to do experiments. Labs required a certain level of dexterity, and I was (and am) hopelessly antidextrous. I'm not clumsy on a grand scale. I don't stumble over my own feet or trip over thresholds, bang my head on the car door frame or my shins on the coffee table. But when it comes to fine motor abilities, I live on the flat and lonely end of the bell curve. All I can do is look longingly up the slope at the ordinary folk who can paint window trim, butter layers of phyllo dough, and braid their daughters' hair neatly. As for the musicians, craftsmen, and visual artists who live on the plain over the far side of the hill, I am awestruck. Such control and coordination of hand, eye, and mind are inconceivable to me. No more can I imagine how a fish feels when it "breathes" water through its gills. I'm beyond all thumbs; my digits feel as though they belong to some other person. It's as if I'm the substitute teacher to a class of ten: My students may do what I ask of them, but in a maddeningly quarter-hearted sort of way. They comply, but only just enough to avoid being sent to the principal's office.

I first became aware of my deficiency when I was ten. My fourth-grade teacher was Miss Sosner, she of the flowered dresses and coordinating pastel cardigans and the Snow White looks. I adored her, not only for the perfection of her person, but because when she found me surreptitiously reading a library book, half on my lap and half inside the metal book box beneath my desk, she merely told me to put it away. (My third-grade teacher had regularly confiscated my books. Fortunately, the library had plenty more.) With Miss Sosner as our guide, our class was to put aside childish printing and learn to write in grown-up cursive, and I was eager to make this transition. Several afternoons each week, Miss Sosner passed out sheets of a slightly aqua-tinted paper printed with rows of solid and dotted lines. After demonstrating on the blackboard the technique for making the letter of the day and how to link it to other letters, we would practice, making line after line of letters and words. I recall finding this hard going, and probably had an inkling that I wasn't shining in the subject. Still, I was mortified when one afternoon, my heroine stooped low in front of my desk and quietly told me that it would be fine if I were to stick with printing for the rest of the year. No other child was so excused, and I understood that I was a hopeless case.

I never did master cursive (although I have a lovely recurring dream in which I effortlessly pen an elaborate, eighteenth-century script adorned with filigrees and swooping furbelows.) I soon realized my ineptitude had other ramifications. I couldn't weave a lanyard without kinks, knit at all, frost a cake without getting crumbs in the frosting, or make a "bridge" while shuffling cards. In my hands, a zipper inevitably snagged the jacket lining, and I resigned myself to wiggling into and out of outerwear left permanently half zipped. Later, I discovered that I couldn't paint my fingernails without also painting my fingers or apply eyeliner so that both eyes matched. I used only clear lip gloss. All of which is to say that in high school chemistry, when it came time to use a pipette to titrate a fluid, I was in trouble. Where one drop was needed, two or three escaped, and I missed the equilibrium point. How did others reliably add water to a flask so that the meniscus just touched the milliliter line? Mine was always just a tick above or below. Measuring a powder onto a delicate balance with what seemed like a toothpick left me two steps behind in experiments. Even connecting the spike of the gas outlet to the base of the Bunsen burner with a rubber hose was a bit of a challenge.

In biology, my fingers rebelled at cutting a slice of onion thin enough to look at under a microscope. In fact, the microscope turned out to be an instrument of torture. I struggled with an early assignment to investigate the protozoa that live in pond water. Preparing a slide, which involved using a pin to position a coverslip (a little glass square about as corporeal as a whisper) on top of a bead of water without trapping any air bubbles beneath it, was an accomplishment. I then looked through the eyepiece and searched for something alive, slowly pushing the slide back and forth with my left hand while—in theory—simultaneously turning the focus knob with my right. For a person who really and truly cannot pat her head and rub her tummy at the same time, this was an excruciating exercise. It turned out that everything under the microscope was upside down and reversed, and I couldn't manage to move the slide smoothly in what felt like the wrong direction. When a tiny creature did come into view, my refractory hand couldn't stop the slide quickly enough, so I shot right past my prey, then struggled to find it again. In addition, I discovered that to a paramecium the narrow layer of water beneath the coverslip is a deep sea. When it swam up or down, it grew blurry and then quickly disappeared from sight. In desperate pursuit, I turned the focus knob so far I drilled the objective lens into the slide, cracking the coverslip, a major laboratory faux pas. When I finished the course, I was relieved to think I would never put my hands on the device again.

Recently, however, in an attempt to experience the world of the early botanists, I met up with microscopes again. The University of Notre Dame has a collection of antique and antique-replica microscopes from the late seventeenth and early eighteenth centuries, and Professor Phillip Sloan, now emeritus, was kind enough to let me use them (under his careful supervision). One of the earliest ones I tried was an English tripod instrument. Its barrel was composed of two pasteboard tubes, one covered in red vellum and the other in green, both intricately tooled with gold leaf. The green tube was slightly smaller so it slid in and out of the red one for gross focus. The larger tube had a wooden eyepiece to look through. The microscope had no light source, so I used it, as it was meant to be used, like a telescope, and pointed it out a window at the bright sky. I had plucked a leaf on my way, and held it on the wooden platform, which had a dime-sized hole in it, near the objective lens at the far end of the smaller barrel. For fine focus, the smaller barrel had a large wooden screw on it, which I turned. Mostly what I saw was green murk, but Dr. Sloan assured me that fault lay in the instruments of the time, not my hands.

An early English tripod microscope.

It didn't surprise me, therefore, to learn that while Galileo's 1609 telescope had been an instant hit and was copied widely, his microscope of the same year, much like the one I had tried, was greeted with little enthusiasm. Not only did early users have to cope with an image that started off "dark and gloomy" and dimmed with every passing cloud, but the view was distorted by the poor quality of glass. Even when they managed to get a specimen in focus, the image had blue and red fringes and blurred edges due to the chromatic and spherical aberrations of the lenses.

Telescopes, moreover, had a practical value, especially as a marine spyglass. Sailors could see the ruffled water over a sandbar in time to change course or spot a merchant ship they hoped to pirate. Astronomy had rapidly become a popular interest of men of good education. In 1665, Samuel Pepys (pronounced *Peeps*), a Royal Navy administrator and enthusiast of natural philosophy whose diaries would bring him posthumous fame, bought a twelve-foot telescope to look at the moon and Jupiter from his house. He already had a "pocket perspective," an early form of binocular, that he brought to church to use, surreptitiously, for "the great pleasure of seeing and gazing at a great many very fine women." But a microscope neither saved your life nor made you a fortune nor served your erotic interests.

Besides, when you looked through a telescope, the unknown became more comprehensible and even familiar. The speck on the horizon became a ship; the landscape of the splotchy moon turned out to look something like Earth's with mountains and valleys. Look into a microscope, however, and the opposite occurred. The familiar became unfamiliar. Red blood looked gray. A green leaf turned into pond scum. The more powerful the lens, the more the view might mystify. Magnify enough and a specimen lost all connection with the known world, becoming only lines and circles and squiggles and empty spaces. Look at a specimen through two different microscopes or by the light of a candle instead of the light from a window, and the object transformed. As Robert Hooke noted in frustration, it was "exceedingly difficult . . . to distinguish between a prominancy and a depression, between a shadow and a black stain, or a reflection and a whiteness." The surface of a fly's eye might look like a lattice drilled with holes or a solid surface covered in golden nails. What truth, if any truth, was in a microscope?

Gradually, microscopes' optics improved. European lens grinders used mathematics to design the shape of lenses, rather than relying on trial and error. Glass manufacture improved and lenses became thinner, which allowed more light to pass up the tube to the eye, brightening the view. Adding a third lens, called a field lens, between the objective lens at the bottom of the tube and the ocular lens at the top, widened the narrow vista. Pepys bought an expensive microscope in London in 1664—"a curious bauble," he called it, an indication of how rare still it was—but had trouble using his new toy. The wonders of the very small remained hidden from all but a handful of specialists.

One of the earliest to master it would be the scientific genius Robert Hooke. No one would have put money on the thirteen-year-old Robert to become one of the stars of Enlightenment science. His father had been an Anglican curate on the Isle of Wight and a royalist sympathizer during the civil war that started in 1642. At his death in 1648, the year before Charles I was beheaded and the Puritan Commonwealth declared, all he was able to leave for his thirteen-year-old son was a paltry £40, a chest, and a collection of books. How exactly the boy made his way from his island home to the prestigious Westminster School in London and into the care of its celebrated headmaster, Dr. Richard Busby, is unclear. Possibly a family friend had a connection with the royalist Busby. (How Busby kept his position under the Commonwealth is another small mystery.) In any case, Robert understood that whatever gentle wave had washed him to the quiet shores of Westminster, he would have to work like a demon to maintain his purchase there. The slender, pale boy with protuberant gray eyes and curly brown hair immediately impressed Busby by memorizing the first six books of Euclid in a week. The headmaster recognized not only the boy's mathematical gifts, but also his artistic talent and manual ingenuity. Knowing that the ecclesiastical career his charge had once expected was foreclosed and that he would need to earn his way, Busby steered him toward the study of mechanics. Perhaps the boy might make his living as a scientific-instrument maker. Maybe he could be a technical assistant cum secretary cum children's tutor to one of the wealthy Englishmen who were just then taking up tinkering in home laboratories.

In 1653, when Hooke was eighteen, Busby managed to get him a place as a scholarship student at his former college, Christ Church, at Oxford. Hooke fell in with a circle of mathematically and scientifically minded young men who gathered around the warden of Wadham College, the charismatic Dr. John Wilkins. Wilkins was a tolerant man in grossly intolerant times, a man chiefly interested in finding and supporting others who he believed would advance the new "experimental philosophy" that animated certain Englishmen of the day. Hooke, as convinced as any of his new friends by the idea that nature must be "put to torture" to give up her secrets, was soon helping members of Wilkins's circle run their investigations.

When Hooke finished at Christ Church, Wilkins helped him get a full-time position as an "operative" to Robert Boyle, a wealthy young Irish aristocrat who had dedicated himself to scientific research. Boyle was then embarking on a study of the physical properties, particularly the "springiness," of air. That springiness, by which he meant compressed air's tendency to rebound, he planned to investigate with an air pump, or what we would call a vacuum pump. Otto von Guericke had invented a primitive version about 1650, but Boyle needed a better one. It was Hooke who managed to construct the device with its pistons, cylinders, and valves coated in "Sallad Oyl" as a sealant, and who was the only one who could get the temperamental machine to work consistently. It was also Hooke who had the mathematics to turn the resulting data into "Boyle's law," which holds that there is an inverse relationship between pressure and the volume of gases. In short order, Hooke had become Boyle's indispensable employee, as well as friend.

In 1660, a dozen scientifically minded gentlemen, including Boyle, Boyle's close friend Christopher Wren, and others from Oxford, gathered in London to found an organization dedicated to the advancement of "Physico-Mathematicall Experimentall Learning." The organization's subjects of inquiry were to be "Physick, Anatomy, Geometry, Astronomy, Navigations, Staticks, Mechanicks, and natural Experiments"; its goal was to discover the laws governing the material world. Its members would consult only their experience, experiment, and observation. There was to be nothing of God or politics (and the membership was indeed a mix of Anglicans and dissenters, royalists and parliamentarians). The brash spirit of the enterprise was emblazoned on the Society's coat of arms: *Nullius in Verba,* or "Take no man's word for it."

The monarchy was restored in May 1660, and in November Charles II chartered the organization, making it the Royal Society of London for Improving Natural Knowledge by Experiment, or more succinctly, the Royal Society. The Crown provided no financial support, but its royal charter gave it prestige and, better, the ability to publish without the standard government-granted license. That meant the group could disseminate information quickly and without censorship, a rare privilege at a time before governments acknowledged a right to a free press. The mathematician Viscount William Brouncker was elected president; Dr. Wilkins and Henry Oldenburg, a former German diplomat employed by Boyle as an amanuensis, were named co-secretaries. The Fellows, all gentlemen with independent incomes, would meet weekly. In November 1662, thanks to Boyle, Hooke was hired as Curator of Experiments, a position that required him to assist in four demonstrations of the Fellows' experiments each week.

The early years of the Royal Society reflected the precarious state of scientific inquiry, teetering between a medieval worldview and modernity. The Fellows, more than a hundred strong, read their own papers, heard letters from some thirty foreign correspondents, hypothesized endlessly, and proposed research projects both sensible and bizarre. One of the founders, Sir Robert Moray, submitted a paper in which he discussed the shells he had seen in Scotland adhering to trees where, he said, they harbored little birds. Others reported that young vipers arose out of powdered viper livers and lungs, and that illnesses could be cured by magnetism. George Villiers, the second Duke of Buckingham, submitted what he contended was the horn of a unicorn. But Boyle, Wren, mathematicians Isaac Barrow and John Wallis (who gave us the googol), ichthyologist

Francis Willoughby, the chemist Thomas Willis, naturalist John Ray, and many others were productive scientists.* Richard Lower, a physician, and Edmund King transfused blood between greyhounds, mastiffs, and sheep, and ultimately between a sheep and one young and very inebriated pub habitué. As a company, they witnessed Boyle's gas experiments, Huygens's pendulum experiments, and Mariotte's investigations of the eye's blind spot. Hoping that their work would have some practical and remunerative results—which was the king's primary reason for chartering the organization—they also tested new brewing methods, compared the best soils and clays for brick-making, and investigated whether adding lime to soil would boost its fertility. Hooke took part in many of these experiments, as well as others involving carriages, fountains, clocks, lenses, chemicals, barometers, hygrometers, and magnets. The job required enormous energy, organizational ability, abstract knowledge, and mechanical skills, as well as considerable tact, given that every virtuoso, especially Boyle who employed him independently, considered his own work to be all-important. To their credit, everyone understood that Hooke was essential to the success of the Royal Society, and elected him a full member in June 1663 while continuing his salary.

Somehow, Hooke managed to find time for his own work. Since the mid-1650s, he and Wren had taken an interest in the microscope and what could be seen beneath

---

* The term *scientist* would not be coined until 1834. *Natural philosopher* remained the term of art until then.

its lenses. By 1661, Hooke had improved its design significantly by engineering a "scotoscope," which consisted of an oil lamp, a glass globe filled with water, and a convex lens that concentrated and directed the lamplight onto a specimen.

He drew what he saw, and in March 1663, the Society entreated him to continue his microscopial observations and to demonstrate the views at the Society's weekly meetings. One year later, it authorized publication of Hooke's *Micrographia, or Some Physiological Descriptions of Minute Bodies Made by Magnifying Glasses with Observations and Inquiries Thereupon.*

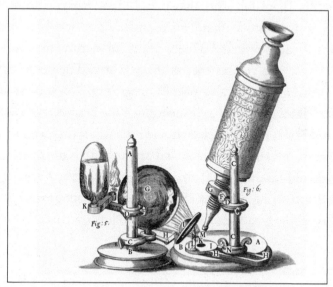

From *Micrographia,* the microscope Robert Hooke used.
Next to the instrument is Hooke's scotoscope. The light
of an oil lamp (K) passes through a water-filled globe (G).
A lens (I) further concentrates light on the specimen.

The large-format book was a sensation for its detailed and beautifully executed illustrations of magnified specimens, ranging from a razor's edge to a snowflake to a louse. Hooke was the first to show the world that a snail has teeth, a bee's stinger has barbs at its tip, and a fly has "feathers" in the middle of its face. In a thin slice of cork, he saw it was composed of "little boxes or cells" (as in the cells of monks) without passageways between them. If Galileo lifted his readers to the stars, Hooke shrunk them to the size of a pinhead, turning a fly into a monster and a nettle leaf into a nightmarish landscape.

The illustrations were not what Hooke actually saw at the bottom of his microscope—the area of focus was too small for that—but were the sum of many partial views artfully crafted into a whole. *Micrographia* was utterly unprecedented, and revealed a new realm of nature that might be as vast and rich as the ordinary world. The perspective was entrancing but also deeply unsettling: No ancient and revered authority had ever mentioned the existence of such a realm. Today, we're so accustomed to such sights, and those far more magnified, that it is hard to fully appreciate the shock and fascination of the drawings when they were first published. If a Mars Rover were to send back photographs of microscopic creatures we would be no less transfixed.

From *Micrographia,* the stinging nettle with sacs of fluid at the base of its needles.

The book was a huge success, print runs sold out repeatedly, and the finances of the straitened Society benefited considerably. Pepys bought a copy immediately on its release and stayed up until two o'clock in the morning reading it, declaring it "the most ingenious book that I ever read in my life." Other London gentlemen, Pepys noted in his diary, rushed out to buy their own microscopes so they could go "microscoping" and see the wonders for themselves.*

For the most part, Hooke was happy to be a tourist of the microscopic world, sending home the most marvelous picture postcards and writing detailed, lively descriptions of exactly how everything looked. Despite his deep understanding of the elaborate mechanisms of machines, he did not try to take apart the insects and plants he viewed. He called himself a "mechanical philosopher" and his greatest love was for the way the inanimate world worked. Chemistry and combustion interested him, and he would invent a clock, the sash window, the universal joint; write a book

---

* For most readers, the minuscule realm was a marvel. However, the poet laureate of the day, Thomas Shadwell, called Hooke "a Sot, that has spent 2000 £ in Microscopes, to find out the nature of Eels in Vinegar, Mites in Cheese, and the Blue of Plums."

about comets; and become a surveyor and an architect, while continuing, for forty-one years, to demonstrate experiments at the Society. With Wren he would design most of the major Royal and City buildings after the Great Fire of London in 1666. But he didn't think to use his microscope to look inside his living specimens—the louse, the fly, and the nettle—to see what internal structures they had or how they operated. Instead, he spoke of an *anima* in plants, a kind of spirit that makes them "useful to the great work or function they are to perform." Aristotle would have said as much.

But if Hooke didn't think to explore the inside of plants, others did.

four

The Persecuted Professor

It is late January in the year 1672, and Marcello Malpighi, forty-three-year-old professor of practical medicine, is sitting in the front row of the anatomy amphitheater at the University of Bologna, attending what is known as the public anatomy lesson. Below him, on the white marble slab at the center of the room, is a human body, gleaming red and partially eviscerated. A man in a bloody coat stands at the side of the slab. He is the dissector, and after excavating an organ he holds it up, turning around to exhibit it to the audience. The lecturer, a medical professor in a black gown and fur collar, stands at the far end of the amphitheater in the cathedra, an elevated and canopied pulpit, and reads from one of the texts of Galen, the great Roman physician, to explain to the audience what it is seeing. There are many other professors in the room, and they pepper the lecturer with questions and objec-

tions. But Malpighi, the world's most knowledgeable anatomist, has said nothing since the proceedings began. His dark
eyes are impassive, his pale face framed by shoulder-length
black hair and sectioned by a narrow black mustache and
a vertical line of beard. From time to time, he takes notes,
not on the bloody organs that the dissector raises, but on the
other professors' comments.

This year's dissection is drawing an even larger crowd than
usual: The subject is a woman just postpartum, quite a rarity.
The amphitheater, the pride of Bologna and the university, is
designed for drama. The wooden ceiling is deeply coffered
and ornate with carved medallions, rosettes, and Latin-
inscribed scrolls. At its recessed center, a languid Apollo
hovers over the proceedings below. The walls, hung with
red damask for the occasion, have deep niches that shelter
life-sized statues of twelve classical philosophers and busts
of renowned Bolognese medical professors. Candles in wall
sconces weakly illuminate the room. Wax torches at the head
and feet of the cadaver cast a fiercer light.

Around the railing that encircles the slab, medical students crowd three-deep. Behind them, benches that look like
choir stalls rise steeply in rows. Government officials in full
scarlet robes are seated at the far end of the amphitheater.
The cardinal legate—Bologna is a papal state—is dressed in
lace, a red skullcap, and red stole and university officials in
black gowns occupy the seats nearest the lecturer. Students,
professors, clergymen, gentlemen and ladies in their rich
green, yellow, and rose silks with full sleeves and broad lace
collars, and members of the ordinary public, many of them
rowdy masqueraders disguised in the elaborate costumes of

Carnival, cram the remaining seats on the two longer sides. In fact, the public anatomy is timed to coincide with the holiday, so as many people as possible, including foreigners who arrive especially for the event, can be dazzled by the building, impressed by the intellectual caliber of the university, possibly educated, and certainly entertained.

Corpses for the anatomy lesson are often hard to come by. Regulations prohibit the use of the body of anyone who had resided within thirty miles of the city. Sometimes a public execution is delayed to provide a fresh subject, and occasionally a doomed convict benefits by having his sentence reduced from a protracted drawing-and-quartering, which produces a cadaver unsuited for teaching, to a quicker hanging. However, at least one unfortunate has had his sentence changed from life imprisonment to hanging to accommodate the annual event. (Lest you think these Bolognese are heartless, the anatomy professor pays for masses to be said for the departed soul.) In any event, the scarcity of cadavers means one has to last for as long as two weeks. Fortunately, Bologna is cold in mid-winter. The intestines and stomach, the organs most likely to stink, are always the first to be removed and dissected. Still, by the end of the session the smells can be as vivid as the sights.

Professor Malpighi isn't bothered by the odors. In fact, this particular corpse and some of its organs will soon find their way to his house in town, where he and select students will continue to scrutinize them. They will inspect the intricate structure of the veins, arteries, and nerves of the ovaries and the connections between the uterus and the remains of the placenta. Painstaking, microscopic dissections undertaken privately have been Malpighi's preoccupation for ten years. Whatever time he can spare from teaching and tending to patients, he spends in his home laboratory. Although he continues to attend the public sessions—it would be grossly impolitic not to—it is no secret that he finds them ridiculous. How can a lecturer reading from Galen about organs that a butcher of a dissector has wrenched from a body teach anything meaningful? It is not unusual for a dissector to be holding up one organ while the lecturer, standing half the length of the room away, is reading about another. For Malpighi, the public anatomies, like his medical practice, are disagreeable necessities, distractions from his real work.

This particular session is worse than a waste of his time. The lecturer suffering in the cathedra, Giovanni Carlo Lanzi Paltroni, is his friend and former student and now a faculty member. Until now, Paltroni had been able to avoid a turn in the *rotuli,* the annual rotation of anatomy professors who lecture publicly. The cost of taking on the lecture series—in addition to paying for masses for the deceased, a first-time lecturer has to buy expensive gifts for officials and students—has been a disincentive. But worse has been the prospect of the *disputatio,* the verbal thrust-and-parry phase between the lecturer and other professors and students, which is a highlight of the event. The government officials in attendance determine the lecturer's salary, and his performance will weigh heavily in their calculations. Paltroni has feared, with good reason, an all-out attack by a powerful group of the university's most prominent medical professors, including Professors Giovanni Sbaraglia and Paolo Mini.

Paltroni's problem is his friendship with Malpighi. Malpighi is a mild-mannered and courteous if rather gloomy man, a conscientious mentor to his medical students, and an artist and connoisseur of Italian painting. Nonetheless, he is despised by most of his colleagues. The Bologna medical faculty members are archconservatives in the world of medicine, strict adherents of the ancient Greek and Arab medicine that the Catholic Church supports. Recently, they have established a new rule: All new medical professors must swear to the doctrines of Aristotle, Hippocrates, and Galen, and "never allow their principles and conclusions to be overturned or destroyed by anyone." One writer who is decidedly not on the curriculum is Vesalius, the sixteenth-century Belgian physician and anatomist who exposed more than two hundred errors in Galen's treatises.* The purpose of the autopsy, from the "obscurantists'" point of view, is not to discover new information, but to demonstrate the accuracy of ancient knowledge. Any disparities between Galen's description of human organs and what one can actually see under the glare of the torches can only be *apparent* differences, differences that the lecturer is bound to resolve in Galen's favor.

Malpighi's zeal for anatomy arouses suspicion among

---

* In early Christian Rome it had been forbidden to dissect humans, so Galen used Barbary apes and other animals, assuming their anatomy was essentially the same as humans'. When Vesalius attended the University of Paris in the 1530s, dissecting executed criminals for the purpose of teaching Galen was acceptable, although as a mere student Vesalius wasn't given that opportunity. Undeterred, he stole bodies from gallows and graves outside the city walls. His 1543 masterwork, *De Humani Coproris Fabrica,* portrayed the anatomy of the human skeleton and musculature with far greater accuracy than ever before—and revealed that Galen hadn't worked from human bodies. Following publication, he was so harassed and slandered by the Catholic Church and various university officials that he burned his unpublished work, renounced further scientific endeavors, and restricted himself to service as court physician, first to Emperor Charles V and later to King Philip II of Spain. He died at age fifty during a pilgrimage to Jerusalem.

many of his fellow professors. To his opponents this intensive exploration of the body's organs is a repudiation of
the ancients' work. Illness, according to Hippocrates and
affirmed by Galen, is caused by an excess or deficiency of one
or more of the four "humors": blood, phlegm, yellow bile,
and black bile. To balance the humors, medical treatments
involve bloodletting and drugs (usually plant-based) that
induce vomiting, diarrhea, urination, and sweating. Other
treatments work by offsetting the symptoms of an excess or
deficiency of a humor, like drinking "cooling asses' milk" to
counter fever brought on by overheated blood.

Another strike against Malpighi is that he relies on a
microscope—a small, upright model created by Italian instrument maker Divini—in his work of prising apart organs and
tissues. In one, typical, six-month period, Malpighi wrote up
and illustrated minute dissections of pregnant and nongravid
cows; two girls, one a pregnant eighteen-year-old and the
other a nineteen-year-old virgin; the penile glands of a dog;
the gallbladder and bile ducts of a snake; the fleshy fibers
in a ray's gills and a spiral valve in its intestines; the uterus
and spleen of a dogfish; the skin on a dog's pads; a squid; the
reproductive organs, liver, and kidneys of a mouse; the eye of
a cow; the tongue of a horse; the skin of a human hand; and
the paws of a mole. Little escapes his scalpel. Several of his
discoveries will be immortalized in modern medical terminology, including the *malpighian layer* of skin and the *malpighian bodies* of the kidney. His most notable microscopial
discovery, however, is the existence of capillaries, the minute
blood vessels connecting arteries that carry blood from the
heart to veins that return it. Capillaries are the missing links,

literal and figurative, that clinch William Harvey's theory that blood circulates, and is not, as Galen believed, used up in the body as food. Malpighi's detailed reports, thoroughly illustrated, on the microstructure of human lungs, tongue, brain, skin, and many other organs have been published in Latin in Italy between 1661 and 1667.

The Bologna obscurantists do not deny the truth of their colleague's findings, but dismiss them nonetheless. It is not possible, they say, that structures so tiny they can only be seen through a microscope could have anything to do with how a body operates. No more would detailing the delicate embroidery on a handkerchief tell you anything essential about how a handkerchief functions. Professor Sbaraglia points out that although Galen promoted the investigation of the shapes, positions, and connections of major organs, he believed that examining the smaller parts was useless. Galenic cures have nothing to do with remedying the malfunction of organs, which is why Professor Mini declares "anatomy does not contribute to medicine." Besides, Galen, the world's greatest physician, did not use microscopes, ergo, microscopial information is irrelevant to medicine. In the years ahead, Mini will urge students to stop dissecting since it is "performed only by persons of little talent and little brain."

Malpighi participated in the *disputatio* when he was younger, but now he often attends without wearing his medical robes and simply listens. (Because his family had not lived in Bologna for the requisite number of generations, he was technically ineligible to lecture in the *rotuli,* although if his colleagues hadn't detested him, no doubt an exception would have been made.) Today, he hears Paltroni discuss the papillae, which are the tiny, sensory bumps on the cutis, an inner layer of skin that Malpighi had discovered. Paltroni immediately falls into trouble. One professor argues that "there is no cutis in the glans penis and [yet] it is sensitive; ergo, the cutis is not the external organ of touch." Sbaraglia declares that "the brain is the organ that gives indications of tangible qualities; ergo, the cutis is not the organ of touch." As Paltroni explains the function of other organs that Malpighi has written about, the professors challenge him scornfully. How ludicrous to think that kidneys filter blood or that muscles move by contraction rather than by "inclination."

There has long been an additional component, a personal one, to Sbaraglia's animus for Malpighi. In 1659, Malpighi's brother, Bartolommeo, stabbed Sbaraglia's older brother to death during an argument on the dark streets of Bologna. Salt in the stiletto wound: Although Bartolommeo was at first condemned to death and all his property was to be confiscated, eighteen months later the court pardoned him and required a payment of only ninety-nine ducats. At the time of Paltroni's lecture, Sbaraglia and company's antagonism has been spurred by their colleague's burgeoning international fame. Henry Oldenburg, secretary of the Royal Society, having read Malpighi's work on the anatomy of lungs, had written to ask if the author had any works on any other subject—animal, mineral, or vegetable—he could send. The silkworm, he mentioned, was of particular interest to the Society. Malpighi, flattered and delighted to have contact with sympathetic intellects, immediately took up the subject, and spent the next year anatomizing all stages of the *bombyce*, from larva to moth. His sixty-thousand-word, illustrated opus, the first-ever exploration of a dissected invertebrate, dazzled the Society members. They had been amazed by Robert Hooke's close-up portrait of a flea, but this work was vastly more informative. Aristotle had written—and no one had looked for any evidence to the contrary—that insects have no internal organs, except perhaps a stomach and a gut, and that they didn't breathe. Malpighi demonstrated that the interiors of insects are filled with organs, including tracheae that have openings in their abdomens and conduct air throughout their bodies, as well as tubes that conduct fluids and emit urine. His work shattered the ancient model. The Society immediately voted to publish his *Dissertatio Epistolica de Bombyce* and make its author a Fellow.

Still, of all the possible subjects that Oldenburg offered, why had Malpighi chosen an insect? No doubt he had been eager to make sure his correspondent would be pleased with his work, but the dissection of a silkworm also happened to fit his own research agenda. For all the information he had gathered in recent years on the anatomy of human and animal organs, he felt he had made little progress in understanding how they worked. Examining the intricacies of their insides did not mean understanding their "connection, movement, and use." He was disappointed that his investigations had added nothing to the practice of healing, a fact that seemed to give substance to his colleagues' charges. He chose to analyze the silkworm because, in part, he hoped it would prove to be a simpler, more comprehensible model of an animal.

To his disappointment, his investigation of silkworm anatomy shed no light on human anatomy or illness. The connections among organs were still obscure, and before he finished *de Bombyce*, he decided he needed to try even simpler beings. He had been pondering plants as analogues for animals for a number of years. The hollow tubes in a broken stem of a chestnut tree reminded him of the tracheae of animals, and he thought the tough fibers of the stems might reveal the secrets of bones and growth. In 1668, he began spending most of his time at his country property outside Bologna, where he had an endless supply of plants. Despite having spent that spring sick in bed—he suffered from kidney stones and recurrent fevers, probably caused by the malaria endemic in the region—on November 1, 1671, Malpighi sent a preliminary study to Oldenburg on the anatomy of plants. "If you tell me that my work is superfluous," he wrote in his cover letter, "I shall permit my body, perpetually exhausted with sickness, to rest. But if you consider it not entirely useless, I shall spend the rest of my life perfecting it."

Oldenburg responded immediately, thanking him for what he titled *Anatome plantarum idea*. The Society has "embraced [the work] with greatest pleasure," he reported, and urged him to continue his research "without hesitation" and to send an enlarged and illustrated manuscript. Malpighi received this letter three months later, at the end of January 1672, just at the time he was sitting in the anatomy theater enduring the attacks on Paltroni. "To the kindness of the Royal Society and your own good offices," he replied, "I acknowledge an obligation so great that I can find no words to express it." Harassed by his colleagues and ill, he abandoned Bologna for the country, and promised to devote himself to finishing the manuscript, confident that he would make a unique contribution to science and medicine.

five

Inside a Plant

What Oldenburg did not tell Malpighi in that letter was that the Royal Society had already received another excellent preliminary work on the anatomy of plants, submitted by one Nehemiah Grew, a twenty-nine-year-old English country medical practitioner. Perhaps Oldenburg, a diplomat by training, chose to give his suffering Italian correspondent time to enjoy unalloyed good news. Maybe he thought that if Malpighi started the project, he would be less likely to abandon it, to the detriment of science and the Society. In any case, after dispatching the letter, Oldenburg and other Fellows then had to tend to Grew, who, on learning of the Italian's work, promptly offered to withdraw from the field. Grew was overawed. He didn't even own a microscope, and Malpighi's anatomical studies were renowned.

Nehemiah Grew had grown up in the town of Coventry, about a hundred miles northwest of London. His father, Obadiah, had attended Oxford and was then ordained in the Church of England and granted a living in Coventry. When the Civil War broke out in 1642, Obadiah sided with the parliamentary party, which had a stronghold in the town, and became a leader in the opposition. After the restoration of the monarchy and the authority of the Anglican Church in 1661, his conscience would not allow him to swear the religious oaths required under the Act of Uniformity, and he had to resign his living. (Four years later, such "ejected" ministers were required to live at least five miles from their former parishes, and he had to leave the town altogether.) His son was lucky to have finished his undergraduate degree at Cambridge just before Nonconformists were barred from the national universities.

Nehemiah returned to Coventry to live with his father, but found himself in a difficult spot. His Cambridge education had prepared him for a Nonconformist ministry, but now a clerical career promised only persecution. Other university-educated Nonconformists, including his older half brother, Henry Sampson, were heading for medicine, having discovered that people tend to overlook the fine points of a doctor's beliefs if they like the medical care. Qualifying to practice medicine, though, was another matter. A physician might earn a fine living in London, but the city required a university medical degree to practice. A man didn't need a degree to practice in the countryside, but he would need the approval of the relevant Anglican bishop, so that avenue was closed, too. Then there was the Leiden route. The University of Leiden in the Netherlands was non-sectarian and drew men—including Henry Sampson—from across Europe whose minority religious views, whether Catholic, Protestant, or Jewish, denied them access to their national institutions. (Women were unwanted everywhere.) If Grew had already been competent, he could simply have paid a fee, taken an exam, submitted a paper, and returned with a degree in a few weeks. But he was an utter novice.

Finally, it was possible to practice without a degree or a license. Unlicensed practitioners provided much of the medical care in rural England, and in reality there was little difference between the treatments of the officially sanctioned physicians and unauthorized healers. Both ordered the phlebotomies, "vomits," and purges advised by Galen (whose work had been translated into English); both prescribed the folk remedies made from local herbs. Neither knew how to cure infectious or most other serious diseases. Neither the Galenist's enema nor the healer's salve of chamomile, marigold, and earthworms could cure malaria or rickets.

Nehemiah chose this last option. It must have been a painful choice for a Cambridge-educated man, but these were harsh times. He seems to have informally apprenticed himself and studied on his own, and put out his shingle in Coventry. He accumulated patients, thanks to his kindly, down-to-earth personality and a humble piety that considered nature, including its human inhabitants, as a reflection of God's wisdom. The divinely created human body, he wrote, was able not only "to prevent, but also to cure or mitigate diseases," and so he treated his patients cautiously, giving the body time to work its own solution.* "In most wounds," he wrote, "if kept clean . . . the flesh will glew together with its own native balm." He was dubious about the value of the medications of his time, commenting wryly that "if you turn over an herbal [a book of plant-based remedies], you shall find almost every herb to be good for every disease." In any case, he had found that apothecaries' ingredients were often counterfeit. Expensive "red oyl of scorpions" in apothecaries' shops, he wrote, was nothing more than tinted vegetable oil.

In 1664, having started a small garden by his house, no doubt so he could have a supply of genuine ingredients for medications, he decided to study plants to see if he might add to the storehouse of knowledge "which the best botanicks had left bare and empty." He cut open stems, flowers, and roots, looked carefully—at this period, with only an unaided eye or perhaps a hand lens—and drew what he saw. He was in no hurry and had no thought of publication; he was simply hoping to reveal more of God's magnificent natural world, revering Him by appreciating His works. In 1668, his half brother, by then a well-connected London physician, encouraged him to write up his findings, and two years later, Sampson passed on the work to Oldenburg, who shared it with Wilkins and others. On May 11, 1671, the Royal Society licensed for publication *Anatomy of Plants Begun*, which summarized his preliminary discoveries and included a proposal for a program of research. In November, Dr. Grew (at Sampson's insistence, he had recently gone to Leiden to get

---

* I have modernized Grew's capitalization and punctuation for clarity.

his medical degree) traveled from Coventry to London and was admitted as a Fellow of the Royal Society.

News of Malpighi's similar proposal reached Grew shortly after he returned home, and greatly disconcerted him. Wilkins urged him to consider that there was room for more than one investigator into such a novel subject. (Grew was probably echoing Wilkins when he later wrote to Malpighi that "although [one man] may have no mind to deceive, yet it is more likely for one, than for two, to be deceived.") Wilkins also began to solicit funds so that Grew, who depended on the income of his practice, could pursue his research. With the promise of £50 per year, Grew moved to London. The Royal Society's financial support over the next five years would prove inconsistent and often in arrears, but he worked on. Fortunately, he had a nearly Panglossian optimism and an unshakable commitment to completing his work.

Between 1671 and 1679, the Society published Grew's beautifully illustrated essays on roots, trunks, flowers, fruits, and seed development. It also published the two volumes of Malpighi's *Anatome Plantarum*, the fulfillment of *Anatome Plantarum Idea*. Malpighi had Grew's work translated into Latin, and the two men corresponded amiably if sporadically. (A regular correspondence was difficult; it could take several months for a letter to travel from London to Bologna.) In 1682, the Society issued Grew's masterwork, which incorporated all his essays plus insights he gathered from—and credited to—Malpighi. We can consider Grew's *Anatomy of Plants* to be a compendium of the work of the two men, a sort of textbook or encyclopedia of the subject.

The *Anatomy* was illustrated with dozens of morphological and microscopic drawings. Grew uncovered and drew fine details in flowers, down to the ovules inside ovaries; leaves, down to the tiny holes on their undersides, called *stomata;* roots, down to the microscopic root cap; and details of the internal structures of trunks and stems. He dissected his specimens in a distinctly modern fashion, cutting them longitudinally, obliquely, and transversely to display an astonishing architecture never seen before, and to demonstrate that from species to species and specimen to specimen, that architecture is consistent.

In modern terms, the two men discovered that there are two basic kinds of tissue in plants. One is *parenchyma* (pa-REN-kuh-ma), the spongy, living tissue in leaves, flowers, fruit, and the bulk of a stem. (In a stem, the parenchyma is called the *pith*.) Parenchyma is made of masses of living cells and looks under the microscope, as Grew wrote, like "the froth on beer." The second kind of tissue looks threadlike and is composed of two types: fibers and vessels (meaning tubes or ducts). The fibers include *collenchyma* and *sclerenchyma* whose cells have thickened walls that provide structural support for a plant, like the studs in a wood-frame house. Collenchyma is made of living cells that form, for example, the strings in celery. Sclerenchyma is made of dead cells, and forms the fibers we prize in flax for making linen and in hemp for making rope.

Vessels are strings of cells that have an opening at top and bottom and fit together like sections of pipe, so that they conduct fluids. Grew saw that vessels run vertically through the parenchyma of trunks, stalks, and stems. In trees, they encircle the perimeter of the branch or trunk, just under the bark. In many other nonwoody plants, like maize and wheat, they are randomly distributed throughout the parenchyma. Neither Grew nor Malpighi could see that what they perceived as a single vessel is actually a bundle of vessels, or *vascular bundles.* Each bundle contains two different kinds of conduits, the water-carrying *xylem* and the sugar-carrying *phloem* (which we'll get to momentarily). Looking at the boundary between bark and wood, Grew saw that in woody species, every spring a new layer of sap-carrying vessels, which he called "lympheducts," develops. By the end of the year, the new layer "losing its original softness by degrees . . . turned into a dry and hard ring of perfect wood." Wood, he realized, is "nothing but a mass of antiquated" lympheducts. Each year a new layer grew and lignified, adding a ring to the wood and increasing the tree's girth. What he could not see was that the new growth comes from the cambium, the layer that citrus grafters have to match between the rootstock and bud.

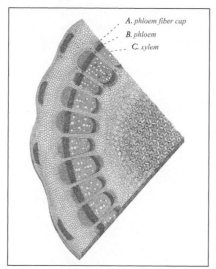

A. *phloem fiber cap*
B. *phloem*
C. *xylem*

Nehemiah Grew's illustration of the stalk of a thistle reveals the basic anatomy of woody plants. In this re-labeled version of the drawing, (A) is the protective phloem fiber cap, (B) is the phloem, and (C) is the xylem. In non-woody (i.e., herbaceous) plants, vascular bundles of xylem and phloem are distributed throughout the parenchyma.

Both men also noted the narrow rows of cells we call *medullary rays*, which run at right angles through the rings of wood from the cambium to the pith. (They couldn't understand their function, but we know the rays transport toxic tannins and other waste products of the living cells at the trunk's periphery into its dead interior where they are safely sequestered. Admire the rich colors of the heartwood of mahogany and walnut? You're looking at the contents of the trees' waste disposal system.) As for the development of new branches, Sir Kenelm Digby, a founding member of the Royal Society, had written authoritatively that juices sent up from the earth accumulate at the top of a tree, where increasing pressure creates a breach in the bark, and "so a new piece . . . is thrust out and begins on the sides, which we call a Branch." Grew disproved this theory. When he dissected axillary leaf buds, he revealed a small miracle. Beneath the outer scales, he found the coming spring's leaves, tiny and intricately folded, waiting for the stem growth that would slide them toward the sun.

Brussels sprouts are axillary leaf buds, meaning they develop in the angle between the leaf stem and the plant stalk. If not picked, a Brussels sprout will grow into a branch and develop leaves and flowers. (When you eat a Brussels sprout, you are eating, botanically speaking, an embryonic branch.)

Grew and Malpighi also realized that trunks grow taller and stems grow longer only at their tips, meaning that plant growth is fundamentally different than animal growth. If children grew like plants, every year their fingers would send out new joints from each fingertip. In fact, initials carved by teenage lovers into a sapling's bark will still be at eye level when the tree is sixty feet tall and the lovers are on their second marriages. And it means that if you plant a sapling whose lowest branches emerge only a few feet off the ground, and you want to stroll under the tree one day, you had better cut off the branches because the trunk will never carry them higher.

The two anatomists changed the way people conceived of plants. Until their books appeared, people considered plants to be an amalgam of parts. The relationship among the parts was . . . well, no one imagined there was much of one. A plant was a kind of Mr. Potato-Head, with roots screwed on like shoes, branches attached like arms, and leaves stuck on toward the top like his eyebrows or hat. When they discovered that the same parenchyma makes up roots and leaves and fruits, and that vessels extend from the tip of a tree root, through the center of the root, up the perimeter of a trunk and out its branches, and into the smallest veins of leaves, a plant made sense as an integral whole. No one would improve on their anatomical work for 150 years.

Despite their accomplishments, however, neither man achieved as much as he had hoped at the outset. Grew had proposed an immense program of research in his *Anatomy of Plants Begun*, of which delineating structure was only the first step. He wanted to understand how plants grow; how seeds are formed, germinate, and send out roots; how "the aliment by which a plant is fed is duly prepared [and] conveyed"; and what causes seasonal changes and what makes colors in flowers, among many other questions. Like Malpighi, he expected that understanding the "how" of plants would shed light on the "how" of animals, including man. But, in fact, they were able to explain very little of plant, much less animal, physiology.

Their failure is not surprising. Their microscopes lacked sufficient magnifying power, for one. Their propensity for analogizing from animals to plants (despite the presumption that it was plants that would shed light on animals) led them astray, too. Malpighi thought he saw peristalsis, the wavelike motion that moves food through the human alimentary canal, in the vessels. Expecting to find organs that served as bowels and lungs in plants, Grew thought vessels, new and "antiquated," were responsible for digestion and breathing. The problem was that uncovering the physiology of plants would take more than meticulous observation. It would require experimentation.

PART II

Roots

six

Restless Roots

One July afternoon more than two decades ago, when our oldest two daughters were two and three, I looked out the kitchen window and saw the western sky had turned a strange and inky green. Usually, I enjoy the drama of a summer thunderstorm, and appreciate the reminder that nature is a powerful force even in our tidy suburban landscape. But this was a storm of a different color, and some marrow-deep instinct told me to find a safer refuge. As a patter of rain escalated into a barrage and the wind churned the treetops to a froth, I hoisted Austen on one hip while I guided Anna down the stairs to our unfinished and windowless basement, turning on the lights and closing the door behind me. I sat on the bottom step, my arms hugging my toddlers close, facing away from whatever was out there. The weather moved in with a staggering roar I'd never heard from weather before.

Abruptly, without the usual flicker, the lights went out, and we were in absolute blackness. The staircase, just open treads nailed to two stringers, vibrated: The four horsemen were galloping close by.

Only a few minutes later, the noise began to recede as the storm thundered its way east, and I could see a gray line of light at the bottom of the door. The three of us clambered up the stairs and went into the living room to look out the French doors into the backyard. At first, the rain was coming down too hard to see anything, but as it slackened and a view gradually materialized, I was stunned. Our yard had disappeared. Instead of a lawn, Ted's vegetable garden, and the metal swing set, I was looking into a solid mass of leaves and branches. The broad crown of the Hansons' sixty-foot sweet gum tree had fallen directly toward our house, its topmost branches no farther than six feet from where we stood. And when I opened the front door, I was greeted by another new landscape. Our street was blocked by at least three fallen trees, their limbs tangled with telephone and electrical wires. I could see houses on Oak Lane that I'd never seen before. A neighbor's maroon station wagon had been crushed under a downed branch like an empty Dr Pepper can.

I hiked the neighborhood the next day. Dozens of trees—tulip poplars and maples, especially, but also oaks and other species I didn't recognize—many of which had stood for a hundred years, had been toppled and were lying on the ground pointing east. To everyone's amazement and relief, no one had been injured. The meteorologists' diagnosis was a microburst, which occurs when a column of cold air plunges from the upper atmosphere during a thunderstorm. (When

a microburst hits the Earth, it rolls outward, creating a tsu-
nami of wind powerful enough not only to topple trees but to
dash airplanes to the ground. A tornado sends trees sprawl-
ing every which way, while the wave of air from a microburst
mows them down in one direction.) It would be ten hot and
humid days before we regained power. Chain saws shrieked
and growled for weeks.

Before a tree service came to dismantle the downed sweet
gum in our yard, the girls and I went to the Hansons' to look
at the base of the trunk. At right angles to the ground was a
massive platter of soil from which emerged an ugly tangle of
arthritic-looking roots. I felt faintly embarrassed staring at
the tree's exposed nether region, as if I'd glanced into a hospital room and seen an unconscious patient with his gown
hiked up. Still, I was fascinated by what I saw, or rather what
I didn't see. Where was the giant taproot I expected? In fact,
hardly any roots emerged directly below the trunk, just
where I thought they should be to anchor the towering mass
above. Instead, not only did most of the roots radiate outward from the base of the trunk like some Jurassic tarantula,
they clearly had been no more than a foot or two beneath
earth. Moreover, none of the roots was as substantial as I
imagined; they were closer to the diameter of a man's fore-
arm than a thigh. Of course, there had to be lots of root ends
left in the earth, but they would be even more slender than
what was exposed. I had to wonder: How had this sixty-foot
giant with its huge, wind-catching canopy stood upright in
even a light breeze? I could understand the mechanics of a
banyan tree. It has a massively broad trunk to start with;
then it drops aerial roots from its branches that thicken over

time and prop up those branches. Mangrove trees I'd seen in Florida seem well designed, too. Their roots arc from a spot a few feet up the trunk into the swamp, acting like flying buttresses. But these neighborhood trees were a puzzle.

A good number of our neighbors were clearly also troubled by the uncertain mechanics of roots. In the years following the microburst, I frequently heard the sound of chain saws, and later would notice another gap in our ever-thinning canopy. In most cases, the trees were healthy; it was the homeowners who were worried sick. (So many were taken down that, ten years ago, the neighborhood council passed an "urban forest ordinance" that limits the removal of healthy trees.) We bucked the trend and bought a scarlet oak sapling to replace the sweet gum, and planted it at the back fence line. Our new tree wouldn't be tall enough to pose a danger for decades. I figured I wouldn't have to concern myself with roots for a long time to come.

As it turned out, I had only a few rootless years. My father retired early, and my parents put their house in Baltimore up for sale. To their frustration, no one offered close to their asking price, and after a year, unwilling to be anchored a moment longer to suburban life, they set sail, literally, for parts south. In their wake, they left their real estate agent with my telephone number and me with a notarized power-of-attorney in case a buyer should appear.

Back in the 1950s, when my newlywed parents moved into their newly constructed ranch-style house with its newly seeded lawn, the builder had put in two silver maple saplings in the front yard. The trees, named for the pale gray underside of their green leaves, grew quickly. But although the shade was welcome and the leaves, which shimmered in the lightest breeze, were beautiful, the ground beneath their canopies was a disaster. Silver maples have notoriously extensive and shallow roots. When I was in grade school, the roots had already crumpled the asphalt driveway. (My mother encouraged me, my sister, and our friends to pull up tarry pieces, a fun but extravagantly messy activity, before she had the area seeded.) Year by year, the roots grew farther out into the lawn while simultaneously heaving themselves ever higher out of the soil. It was as if they, like my parents, had plans for getting out of town.

By the time my parents left Baltimore, hardly any grass grew in the front yard. Whatever blades did emerge from the crevasses among the roots couldn't be cut, unless you were willing to go at them with a pair of scissors. No doubt the state of the yard had something to do with the lack of interest in the property. Most suburban buyers wanted a lawn, not a bas-relief of roots. Luckily for my parents, a single man, a sports photographer who traveled frequently and must have considered a yard that didn't need mowing to be an asset, made an acceptable offer.

All was going smoothly with the transaction until the home inspection report came in. It seemed that the sewer line was not draining as it should. A plumber was summoned to investigate. He called to tell me that a tiny silver-maple root tip had found an equally tiny crack, slipped through it, and, having stumbled upon a rich supply of water and fertilizer, prospered mightily, becoming "an eight-foot horsetail" of silver maple roots. A pipe invasion of this sort was not an uncommon occurrence, he said, especially where clay pipes and maples, willows, and sweet gums were concerned. He'd once found the roots of a maple had traveled under a street to invade and clog a neighbor's sewer line fifty feet away.

Naturally, the photographer was not satisfied with a mere snaking of the pipe, and asked for a visual inspection. Right he was: The root that had taken up residence in the pipe had broadened in girth over time, thereby enlarging the original crack into a significant hole. Many hundreds of dollars later, that section of pipe was replaced, and the deal closed.

I now know that the intruder was a "sinker root," a type of root that drops down from lateral roots in search of water. On encountering the condensation on a clay sewer pipe, sinker roots spread across its damp surface. If a tree is lucky, and the homeowner isn't, a sinker finds access to the inside of the pipe. Sinker roots, although occasionally a menace to sewer and water pipes, are an important part of the answer to why tall trees manage to stay upright under most circumstances. As trees mature, their taproots often atrophy. While shallow lateral roots generally do little to anchor trees, some laterals grow obliquely downward. Even better, sinkers grow straight down. (I never saw the sweet gum's sinkers; they are relatively slender and were left in the earth as the tree fell.) Although the trunk and leafy branches of a tree like the sweet gum might weigh five tons, the weight of its oblique laterals and sinkers, together with the vast amount of soil they enclose within their grasp, can weigh many times more. The roots and soil act, in essence, like the bulbous lead keel of a sailboat.

I'd never had any interest in roots, pale and witchy-fingered things that probe and poke the darkness underground. I didn't like to think about them, and was glad they were usually hidden from sight. But gardeners ought to know that, according to the University of Colorado, about 80 percent of all plant problems are really root problems. The beauty of flowers and the bounty of a harvest have almost everything to do with roots.

seven

The Enormous Gourd

Steve Connolly's prize-winning pumpkin, which I have come to see at the New York Botanical Garden, is the third-heaviest pumpkin grown in North America this year. It weighs nearly 1,700 pounds. Picture a caramel-colored Smart car lying on its side, and you have its rough dimensions. Steve's pumpkin has only one door, though. That door, a rectangular, foot-thick section of yellow-orange rind, is sitting on the ground, leaving a hole that is just large enough for a skinny young man to wiggle through. In fact, such a man is inside the pumpkin, and has just poked his torso and dark-haired head out of the hole to place a white plastic bucket on the ground. He is wearing a headlamp and, despite the autumnal chill, a T-shirt. The bucket, the young man tells me, is full of seeds he has scraped from the interior rind. Inside the pumpkin, he further reports, it is very dark, warm—the strong noon sun seems to turn the pumpkin into a solar oven—and wet. Having issued this report, he disappears again into the pumpkin.

If you were to put this pumpkin on wheels and hitch it to a team of white horses, you could ride to the prince's ball in it. There would be room for your fairy godmother, too. But if you're going to go dancing, it had better be tonight. Tomorrow is October 29, and an artist will be turning this pumpkin and the first-place winner (1,810.5 pounds from Minnesota) into the world's largest jack-o'-lanterns.

I am waiting for Steve to arrive, imagining a hefty farmer in denim overalls. Instead, I find myself shaking hands with a slight, pale, bespectacled, and soft-spoken man in his mid-fifties who I discover has a BS in plastics engineering and a job at a large medical company. He is wearing an orange baseball cap and a jacket emblazoned with the emblem—a bright orange pumpkin impressed with a green map of the world—of the Great Pumpkin Commonwealth. The GPC, he explains, is the umbrella organization of some forty regional clubs whose ten thousand members are devoted to "extreme gardening." Extreme gardening refers to the size of the fruits they grow—tomatoes, squash, long gourds, watermelons, and pumpkins—as well as the fervor they bring to their cultivation. They're committed to doing whatever it takes to grow the largest garden fruits the world has ever seen. As Don Langevin, author of the pumpkin growers' bible, *How-to-Grow World Class Giant Pumpkins*, writes: "From April to September and beyond, a serious giant pumpkin grower must dedicate his or her life to his plants."

So, what does it take to grow an orange gourd that weighs

as much as a small car? You have to start with seeds of the *Cucurbita maxima* "Atlantic Giant." This is not the variety you find in late October dotting the fields of pick-your-own farms. You might be able to purchase Atlantic Giant seeds at your local garden center, but you'll probably raise a pumpkin that weighs a mere several hundred pounds. If you're going for the gold, you had better contact prize winners of the last few years and see if they'll sell you a few of theirs. Or, you can try to buy them at auction. "Proven" seeds, those harvested from a prize winner that produced winners in the next season, can set you back many hundreds of dollars. Seeds from the 2010 champion went for $1,200 each.

You'll also need a good part of an average suburban backyard. Each plant requires a thirty-foot by thirty-foot patch, and it's wise to cultivate several plants. Your patch should be completely sunlit because your pumpkin is going to need every available photon. The shade of a tree might stand between you and immortality, and more than a few trees have given their lives to the cause. You'd also do well to live in the "orange zone," which in North America lies between the 40th and 46th parallels, that is, roughly from San Francisco to Vancouver, from Nebraska up to Ontario, and from Pennsylvania through St. John's, Newfoundland. In this region you will find the greatest number of daylight hours paired with the greatest number of frost-free days.

Even so, Steve warns, you will need to start your seeds indoors under grow lights in mid-April; otherwise you'll be behind the serious competition by the time the danger of frost has passed. Don Langevin advises that if you're new to the sport, you ought to start practicing germinating seeds in early March, in order to build your confidence for when G-day arrives. About May 1, it's time to transplant your young plants outside. Competitive growers build a plastic greenhouse the size of a small doghouse over each one. These can get elaborate, with buried heating cables to raise soil temperature to the ideal 85 degrees. When your vines outgrow their enclosures, you might want to build a windbreak of straw bales close by so they won't get twisted or otherwise damaged.

Left to their own devices, your plants' vines will branch and rebranch and then rebranch some more, crisscrossing each other while growing several inches a day into a wrist-thick spaghetti of vinery. The plants will also want to develop lots of leaves, which are as big as serving platters and hover above the vines on two-foot-tall stems. Your vines will also want to produce many small (relatively speaking) fruits that can hide under those leaves. In this way, even if insects and deer graze the leaves and raccoons or groundhogs discover most of the fruits, the likelihood is that one pumpkin will survive to autumn to produce seeds for the next generation.

You, on the other hand, want all the plant's energy going to develop one monster pumpkin, and you want only as many leaves as are needed to supply the energy for that one fruit's growth. As growers often told me, you are not trying to grow a salad, you are trying to grow a pumpkin. A rampant tangle of vines and leaves means some leaves will shade others, and shaded leaves are slacker leaves when it comes to the business of gathering sunlight. So it's your job to go into the patch *every* day and prune, arrange, and stake, the rapidly growing vines so that they conform to your ideal. You want the vines to grow in a pattern that looks, spread out on the ground, like the profile of a Christmas tree. If you had an aerial view, your pumpkin would look like a humongous orange ornament smack in the middle of that tree.

In early June, it is time for you to pollinate the bright yellow female flowers. Pumpkins have separate male and female flowers growing on one plant. It won't matter to the success of this season's fruit which male flower pollinates your female flowers—your pumpkin's genetics were encoded in the seed you planted. But, if your plumpening darling turns out to be a winner, the value of her seeds will be a function of the seeds' parentage. So, many growers cap their emergent female flowers with a plastic cup or a sock—a sort of pumpkin condom—so that some randy flower on some loser vine won't knock up their pedigreed virgins. When the moment is right, you will brush the females' pistils with pollen either from your male flowers (a process called "selfing") or with the pollen from another grower's pedigreed plant.

Two weeks after pollination your ward will be the size of a basketball. Two weeks after that, it will weigh four hundred pounds. (By this time, you ought to have shoveled some sand under your burgeoning fruit to prevent it from rotting on the underside.) According to the *Engineers Guide to Supersizing Pumpkins* at Impactlab.com, the bigger a gourd gets, the more physical stress it experiences, which triggers it to grow even more. "Their weight generates tension, which pulls cells apart and accelerates growth," writes David Hu at the Georgia Institute of Technology. At around 220 pounds, your round pumpkin will start to flatten under the pressure of its own weight, which is why it will ultimately look like a Brobdingnagian orange pouf rather than a humongous orange orb.

Starting in June, you'll need to shelter your protégée from the sun. The danger is that as your pumpkin packs on weight—up to forty pounds a day in midsummer—its sun-toughened skin will crack under the pressure of its rapidly expanding innards. Steve uses a wooden structure covered with a lightweight, white fabric as a sunscreen; others pile on white towels. Still, sometimes pumpkins explode. It's a sad day when it happens, but you know what they say about making omelets.

But for all the drama and action in the pumpkin patch, the most significant factor in growing a record-breaker is not how you care for the giant vines, leaves, and fruit that you see, but how you care for the roots that you don't. Not surprisingly, soil preparation is an obsession for the most competitive giant pumpkin growers. Champion Len Stellpflug adds five cubic yards of manure—weighing several thousand pounds—to his patch near Rochester, New York. The best manure is a subject of intense debate: Some growers swear by chicken manure mixed with sawdust while others add cow, horse, or alpaca poop. Manure happens, but minerals may not. Providing just the right mix is essential. In addition to sixty pounds of kelp and fifty pounds of humic acid, Ron and Pap Wallace added more than 220 pounds of calcium, manganese, potassium, sulfur, and magnesium, with a soupçon (relatively speaking) of 20 Mule Team Borax, and that was all before June. Serious competitors send a leaf sample to be tested for nutrient analysis every two weeks, and amend their soil accordingly.

An Atlantic Giant is about 90 percent water, so getting enough water to the roots is essential. Some growers use drip irrigation systems; others a combination of sprinklers, misters, and hand watering. Deep into a New England summer they may have to add as much as 125 gallons a day to the ground. Some growers warm the water first in a polypropylene outdoor tank. This is not as silly as it sounds: The theory is that cold water sends a signal to the pumpkin that winter is coming and it should stop growing. Other midsummer techniques that growers use, like massaging their tender baby with oil or bathing it with milk, are more suspect.

Ample water and perfect soil are meaningful only if your pumpkin has a healthy, well-developed root system to take advantage of them; the source of big fruits is big roots. This is why, throughout the season, growers dig a little trench ahead of their vines and lightly bury each one as it creeps forward. At every leaf axil (as well as at root tips and throughout the cambium) is undifferentiated cell tissue called meristem. Axillary meristems are capable of forming a bud for either a new leaf, a flower, or a root, depending on what chemical signals they receive. If you bury the axils, they will produce new taproots, which then send out lateral roots that channel more water and nutrients back to your gourd. The laterals, however, are fragile and run shallow, so most growers lay down wood planks around their patch so they won't inadvertently crush these delicate pioneers. But should you spot a man or woman tending their pumpkins on snow skis, don't be surprised. They believe their skis spread their weight and allow them to pass especially lightly over the soil. In Giant-Pumpkin Land, the road to hell is paved with wood indentions.

eight

The Way of All Water

No doubt, eight thousand years ago the farmers at Catal Hüyük in modern-day Turkey understood that roots had something to do with feeding and sustaining their emmer crop. They could hardly have missed the fact that emmer, the antecedent of modern wheat, grows better in dark, loamy areas than sandy ones and that plants in dry soil go limp and die. Ancient Egyptians certainly knew that without the summer flooding of the Nile and a new coating of silt on the land, there would be poor crops in the spring and famine the following year.

How roots gather food from soil and distribute it throughout the aboveground part of the plant was, however, a deep mystery. Aristotle believed roots took in food from the soil indiscriminately, and concluded that whatever that food was, it was ready to be consumed because he found no stomachlike organ in plants. Theophrastus disagreed: Roots could choose what to take in and had a digestive capability. Eighteen hundred years later, the question was still open. One of the few to write on the subject, the renowned Renaissance physician Girolamo Cardano, concluded that digestion does take place in the plant, but not in the roots. The stomach of the plant, too small to be seen, must be at the base of a plant's stem or a tree's trunk, just above the roots.

That the presumed location of the imaginary stomach is equivalent to the position of our stomach and intestines, which lie at the base of our trunks just above our (root-shaped) legs, is no coincidence. Cardano's plants-to-humans analogy was instinctual, but also encouraged by the prevailing paradigm of the Great Chain of Being. The Great Chain was conceived in the ancient world and exquisitely refined in the Renaissance. It was the model of the relationship among all the world's living and nonliving entities. Imagine the Chain hanging from heaven, with perfect God at the top, the angels below Him, Man just below the angels, and, descending in order of their ever-greater imperfection, the animals, birds, sea creatures, plants, minerals, and finally, stones. Philosophers gave considerable thought to the fine gradations of the hierarchy: Robins, for example, were higher on the Chain than sparrows because robins eat worms (as animals eat meat) while sparrows dine only on seeds. Plants stood higher than minerals because, like animals, they could eat and minerals could not. Plants' digestive organs would certainly reside in their lower torsos.

In the seventeenth century, discoveries about human anatomy seemed to shed new light on vegetal alimentation. Until

that time, people believed that human and animal livers produce cool, venous blood while hearts produce warm, arterial
blood, and the two kinds of blood never mix. In 1628, William Harvey proved by experiment that all blood travels in a
continuous circuit from the heart through arteries and back
via veins whose tiny valves prevent it from flowing backward
between pulses. In 1648, Jean Pecquet, a French physician,
found that a milky fluid he named *chyle* leaves the intestines
via a lymph duct that connects to a vessel near the heart.
Chyle then enters the bloodstream. With these two new facts,
a revised theory of human nutrition was born. The heart and
the arteries "cooked" the chyle and distributed this food to the
body through arterial blood. The blood then returned to the
heart, darker and depleted of nutriment, via the veins.*

These discoveries inspired a parallel theory of plant nutrition. What if sap was the equivalent of our chyle-fortified
blood? What if sap picked up raw nutriment from the soil,
and then circulated it continuously throughout the plant's
body? Now, certain observations made sense. People had long
known that if you cut into the outer wood of a tree or sliced
through the stem of a nonwoody plant, rising sap flowed out.
In the 1670s, Malpighi, in conducting one of his few experiments, added new evidence to the discussions of circulation.
He girdled a tree, cutting through only the bark and not into
the wood, and discovered that the bark above the cut swelled
and the bark below the cut died. Clearly, he had interrupted
some kind of downward flow of liquid.

---

* Chyle is actually a combination of lymphatic fluid and essential lipoproteins.
  Transported by blood, the lipoproteins are either stored or metabolized by
  different tissues in the body.

The conclusion was inevitable: Sap flows upward in vessels under the cambium, turns around in the leaves, and seeps down on the outside of the cambium just beneath the bark. Both Grew and Malpighi had seen the vessels that carry fluid upward, which we know as *xylem*. Grew also managed to observe a tracery of tiny "pipes or tubes" on the underside of bark, vessels we now know as *phloem*. There was additional, circumstantial evidence for the circulation of sap. The liquid flowing upward is more forceful and deeper in the tree's body, just as arterial blood is; the return flow is weaker and closer to the tree's "skin," just as venous blood is. Score another one for the model of the Great Chain of Being: Plant sap circulates like animals' blood, just more simply.

The facts didn't all fall into place, however. It was perplexing that sap didn't seem to flow in all seasons. Nor did anyone see capillaries similar to those that Malpighi discovered in animals that connect the upward flow with the downward seepage. (And, although no one seems to have asked the question, what about nonwoody species with the vessels scattered throughout the parenchyma? Their phloem are bundled with xylem in the interior of the stem, and so were invisible to observers in this era.) But the biggest problem with a theory of a vegetal circulatory system was that a plant had no heart, no engine for pushing fluid up and around.

Many solutions to the problem were proposed. Claude Perrault, a French scientist and architect, suggested that as branches swayed in the wind they compressed the sap vessels, pushing the fluids inside. Grew offered that the little "bladders" he saw in the parenchyma were cisterns for sap, which swelled and squeezed fluid into and through the vessels. (Those little bladders are actually living cells that store sugars and have other functions, depending on their location in a plant.) Malpighi thought the "spiral vessels" he saw wrapped around xylem expanded in the heat of day and contracted at night, squeezing the liquid along its way. (The spiral elements he saw are actually thickenings of the xylem walls that help keep them rigid.) Despite the lack of a consensus on what force impelled sap to circulate, no one doubted that circulate it did.

Not until 1715, did someone finally test the proposition. That someone was the Reverend Stephen Hales, a curate in the English village of Teddington. Hales was the youngest son of a baronet in the county of Kent. He earned his M.A. in divinity at Cambridge in 1703, and then took up a fellowship while he waited for the current occupant of his promised curacy to move on to the greenest of all pastures. Without any duties and at loose ends, this intellectually curious young man befriended an undergraduate medical student, and tagged along with him to his classes.

The goal of the university continued to be, as it had always been, to educate the next generation of ministers, lawyers, and doctors. Most experimentation and research in England at that time took place at the Royal Society and in the home laboratories of wealthy amateurs, not in classrooms. But the university was beginning to offer courses beyond the traditional ones in Latin, rhetoric, logic, moral philosophy, ethics, and basic mathematics. A new observatory had been built on the top of the Great Gate at Trinity and the first chemistry laboratory—although the chemistry was not much more than alchemy—had been constructed on the college's bowling green. For the first time, science lectures incorporated demonstrations, a profound pedagogical innovation. Hales attended these, and found instruction in the new subjects of hydrostatics and pneumatics. He dissected dogs and cats and made lead casts of their lungs, watched electrical experiments, repeated Robert Boyles's air-pump experiments, and designed an orrery based on Isaac Newton's understanding of the solar system.

Newton strongly influenced Hales, although the professor had left Cambridge in 1696, the year Hales matriculated, to take over the operation of the Royal Mint. In fact, it was only after Newton's departure that students were taking courses based on his discoveries and his great works, the *Principia*, *Optics*, and *Arithmetica universalis*. When Newton had been on the faculty, only a few students attended his lectures, and the great man spent most of his time working alone on his physico/mathematical theories and his alchemical experiments. (It was said that as he walked the garden paths at Trinity College, cogitating, he drew diagrams in the gravel that everyone reverentially stepped around.) Hales absorbed the Newtonian model of the natural world, the new model that was beginning to supersede the older, Cartesian one.

René Descartes, who died in 1650, had been a revolutionary thinker in his time. He had rejected the ancient and still prevailing belief that natural objects—and even the universe itself—had souls that animated them, which explained why apples fall down to earth and plants grow up. No invisible forces, which to Descartes and his rationalist cohorts meant

occult forces, were needed to explain the motions of matter. Instead, he posited a clocklike universe that once set in
motion by God ran on its own in perpetuity without further
interference on His part. In the Cartesian model, all motion
was communicated from one part of matter to another by
direct contact, the way gears in a clock's workings engage
and turn each other. It was Newton's insight that unseen
forces *do* play a role in determining the motion of matter.
He could not explain how gravity, magnetism, and cohesion
draw and repulse matter across empty space, but he knew the
forces are real, measurable, and, above all, operate according
to mathematical laws.

By the time Rev. Hales took up his curacy in early 1709,
he was thirty-one years old and had become a thoroughgo-
ing Newtonian. Like Grew and many other English scientists of his time, he also saw no contradiction between his
religious beliefs and scientific interests. As he would write
in his first book, "Not only the grandeur of this our solar
system, and the other heavenly bodies, declare the glory of
God, but also the exceeding minuteness of microscopial animals, and of their component Parts." So, it was with a clear
conscience that once installed at Teddington, he launched
himself simultaneously into his ministry and a program of
natural experimentation. Newton had measured the forces
that governed the movement of objects; Hales, more modestly, set out to measure the forces of the circulatory system
of animals.

Teddington was a quiet village with a population of four
hundred about fifteen miles outside London. It was home to
a single church and two inns, and was surrounded by hun-

dreds of acres of open, low-lying fields bordering the Thames River. For many of Hales's parishioners, a scientifically minded curate must have been a curiosity, although if his focus had been astronomy or classifying plants, his passion would have gone with little comment. That his particular scientific interests made a strong impression is not surprising. In one of his early experiments, he tied a struggling white mare to a gate laid on the ground, squatted to cut a hole in her left carotid artery with a pen knife, and, by then surely blood-spattered, inserted the end of a twelve-foot-long glass tube into the hole in order measure the force of her beating heart. While he conducted his small-scale experiments inside his garden laboratory, when a Cambridge-educated curate slaughters in his yard some five dozen oxen, cows, deer, dogs, and sheep by sticking glass tubes in their veins or lungs, word gets around. His penchant for vivisection made him quite a notable character in the neighborhood. Thomas Twining, a minister and classics scholar who lived nearby, would later write of

Green Teddington's serene retreat
For Philosophic studies meet,
Where the good Pastor Stephen Hales
Weighed moisture in a pair of scales,
To lingering death put Mares and Dogs,
And stripped the Skins from living Frogs.

The poet Alexander Pope, who also lived in the area, was quoted as saying of Hales that "[I] always love to see him, he is so worthy and good a man. Yes, he is a very good man,

only I'm sorry he has his hands so much imbued in blood."
Blood, however, was at the core of Hales's research.

Despite his odd and sanguinary interests, the reverend
was popular in the parish. He was blessedly undogmatic,
and his sermons emphasized God's goodness and the
need for ordinary charity. (On the subject of extramari-
tal relations, however, he was strict, requiring those guilty
of adultery and fornication to wear the customary white
robes to church and beg forgiveness from the assembled
congregation.) He generally avoided the doctrinal issues
that had ignited civil war in the previous century and continued to rive the Protestant community. Attendance at St.
Mary's Church increased after his arrival, and the church
had to be enlarged in 1716 to accommodate his growing
congregation.

In 1715 or so, Hales suddenly diverted his attention
from quantifying the flow of blood in arteries and veins
to studying the passage of sap in plants. Like Malpighi, he
may have hoped that the study of plants would illuminate
the workings of animals. Certainly, it was far easier—and
more socially acceptable—to cut up cabbages than cats. In
1718, he read his first paper at the Royal Society, reporting
"a new Experiment upon the Effect of the Sun's warmth in
raising the Sap in trees," and was promptly elected a Fellow. The Society urged him to continue his work, and in
1727 he presented his groundbreaking treatise, titled *Vegetable Staticks.*

Hales used the word *vegetable*, as everyone did, to mean
all of plant life. By *staticks* he meant both measuring in general, but also specifically the measuring of the input and out-

put of fluids, as well as their force and speed. One of the first questions he addressed was whether sap does in fact circulate, like blood in an animal, through a closed system. If it did, he reasoned, it would flow in one direction only, just as blood does.

To test the theory, he cut off a leafy branch of an apple tree, removed the tip of the branch, then put the branch upside down in a glass tube of water. The tube was connected to a denuded branch placed upside down in a bucket of mercury. (See Figure 13 on the next page.) On a sunny day, the leafy branch emptied the tube of water and drew mercury up toward the denuded branch, proving water could pass easily in the opposite direction of its normal flow. Stripping the branch of bark made no difference to the results. Finally, he tried the experiment with a living branch. (See Figure 14 on the next page.) Sap, he correctly concluded, does not circulate like blood.

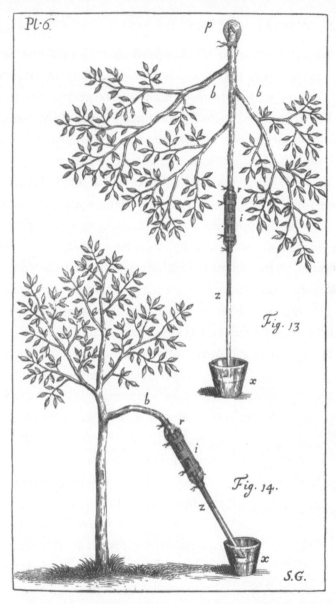

Pl·6.

p

b   *b*

r
i

z

Fig. 13

x

b

r
i

Fig. 14.

z

x

S.G.

In the experiment illustrated in Figure 13, Hales clipped both ends of a leafy branch (b). He denuded another branch (z), and connected (b) and (z), both upside-down, via a water-filled glass tube (i). The other end of (z) he put in a bucket of mercury. He repeated the experiment (Figure 14) with a living tree.

But if it didn't circulate, then where did it go and how did it get there? Hales set up another experiment to investigate those questions. He started with a sunflower growing in a clay pot, then sealed the soil's surface and blocked the hole at the bottom of the pot. He watered the sunflower through a small tube that he corked afterward. By comparing the quantity of water he added to the pot each day with the daily changes in the weight of the pot and plant, he arrived at the amount that disappeared. The water could only have disappeared into the atmosphere. By measuring the surface area of the leaves, stalk, and stems, Hales arrived at the plant's rate of transpiration—"perspiration," as he called it—per square inch. By calculating the surface area of the roots, he also figured the rate at which roots take up water.

He repeated the potted sunflower experiment at different

air temperatures, levels of humidity, and cloud cover, and with lemon trees and other species. Sunny skies and heat, he found, increased transpiration; humidity and clouds decreased it.

He could have concluded that the sun is somehow directly responsible for the rise of sap. Instead, he ran another experiment, testing the "imbibing" capacity of leafy versus leafless branches. While leafy branches

In this experiment with a sunflower, Hales measured the amount of water taken up by roots and lost to the air through the plant's aboveground portion.

pulled up water, leafless ones did not. Sun or no sun, water does not travel up leafless branches. Unlikely as it seemed, he realized it is the almost imperceptible force of evaporation from leaves through the pinprick stomata that moves water from the ground to treetops. In fact, through the power of transpiration, the canopy of a hundred-foot-tall tulip poplar can hoist as much as a hundred gallons of water, weighing a total of eight hundred pounds, per day.

Hales's discoveries raised new questions. If fluid in xylem exited the plant through leaves, then what was the fluid that Malpighi and Grew discovered flowing downward just beneath the bark? Hales missed what is known as phloem sap, the sucrose-laced liquid that flows down from the leaves to the roots for storage. (It also moves sideways and sometimes even upward, taking liquid energy to cells that use it to function.) But phloem sap was impossible for him to tap and measure with his instruments. It moves far more slowly and with far less pressure than xylem sap.

Even today, the pressure of phloem sap is difficult to measure, and plant physiologists have to enlist aphids to do the job for them. Aphids and other garden pests pierce the phloem with a needlelike mouthpart (a *stylet*) in order to route the sugary liquid into their digestive tracts. (Their excretions become food for the unsightly "sooty mold" that gardeners find on leaves.) Modern researchers set aphids to work by placing them on a plant stem where the bugs drill in. The researchers then zap the aphid body with a laser or blow it apart with a jolt of electricity, and measure the flow exiting its disembodied stylet.

I imagine Hales would have been delighted with the

technical ingenuity of this method. After the publication of his two great books, *Vegetable Staticks* followed by *Haemostaticks*, he rechanneled much of his energy into developing technology for the betterment of humanity. In the 1740s, he invented forced-air ventilators that reduced mold in granaries and brought fresh air to prison blocks and the fetid atmosphere belowdecks in merchant ships. The success of his ventilators in reducing illness convinced the Royal Navy to require them in all its ships in 1756, and many English hospitals also adopted them.

In Hales's later years, he investigated the relationship between the mineral encrustations on the bottoms of teakettles and the salubrity of the well water boiled in them. He advised watermen on how to preserve the undersides of their boats; taught housewives to place an inverted teacup at the bottom of pies to keep the syrup from boiling over; tried to poison insects on fruit trees by drilling holes in the trees and pouring in mercury (not a good idea); designed a water conduit system for Teddington; and demonstrated that putting air holes in the outer walls of ground-floor rooms prevents floorboard and joist rot. In an era when well-connected Anglican clergymen often accepted the income from several parishes while providing little or no service to most of them, he declined several opportunities to do so, and lived modestly.

At the age of seventy-four, he was appointed chaplain to the Princess Dowager of Wales, an honor that reflected his scientific achievements, as well as his service to his country. Still, he remained a bit of an odd duck. He had the elderly princess searching the bottom of her teacup for mineral sediments. His neighbor Gilbert White wrote of Hales that his "whole mind seemed replete with experiment, which of course gave a tincture and turn to his conversation, often somewhat peculiar." But his experiments, neatly crafted and reproducible, created the field of plant physiology. No one would significantly advance his work on transpiration for a century.

nine

How to Kill a Hickory

About fifteen years ago, friends up the street built an addition off the back of their house. Using the soil excavated in digging out the addition's foundation, the Carters leveled their yard. Because their small yard sloped sharply down toward their neighbors, they had to build a three-foot-high retaining wall along the joint property line.

A hickory tree nearly forty feet tall stood ten feet from the retaining wall. The Carters and their neighbors enjoyed the tree, not only for its unassuming grace and shimmering golden foliage in autumn, but because its arching boughs extended over their roofs, shading their houses from the sun and no doubt moderating their electricity bills. The Carters knew that loading soil onto the roots of the tree might endanger it, so they had a three-foot by three-foot "tree well" made of railroad ties built around the base of its trunk.

The hickory after its roots were partially buried. A root tip is magnified.

For many years, the hickory prospered. Two summers ago, however, I noticed that its leaves were smaller and sparser than usual, and I idly wondered why. Early this summer and seemingly overnight, its green leaves shriveled and turned a purplish brown. I see that the bark has peeled back from some of its boughs. The tree is now indubitably dead. Although it is still standing as I write this in January, it will be gone by spring.

The Carters assume that their tree is an ill-fated victim of a dry spell last spring. But its death was neither due to drought nor was it inevitable. Instead, its roots slowly and simultaneously starved and suffocated under the soil that was added. The tree well was a well-meant gesture, but it couldn't save the tree. The question is, why not?

Stephen Hales made great strides in explaining how water moves once inside plants, but he didn't explore how water gets into roots in the first place. He did notice, as Malpighi and Grew had decades earlier, a band of superfine, colorless hairs growing just behind the tips of roots. Collectively, these hairs look like microscopic bottlebrushes. All three men dismissed the hairs as unimportant. In fact, they are essential to a plant's survival, responsible for gathering water and nutrients a plant needs from the soil.

A root hair is a single elongated cell of the root's epidermis. At less than a millimeter in length, most root hairs are invisible, which means that their power is in their numbers. A four-month-old rye plant in a twelve-inch pot, for example, has about fourteen billion root hairs. If some modern-day Sisyphus were condemned to lay those hairs end to end, they would stretch from Los Angeles to Boston and nearly back again. Their collective surface vastly increases the area of contact between roots on the one hand and soil, water molecules, and microscopic pockets of air on the other. To enhance their critical water-funneling activity, root hairs excrete a slimy, sugary substance, called *mucigel*, that binds molecules of water and particles of soil to them. Water passes via osmosis (without the expenditure of energy) through a root hair's cell wall first, then through the membrane that lines the cell, and finally through the xylem's endodermis and into the xylem. Plants, including the hickory next door, can only replenish the water they lose via transpiration if they have sufficient root hairs.

Your garden or potted plant is in trouble if more water is evaporating from its leaves than is entering its root hairs. It begins to wilt as water is pulled out of the xylem, then from the spaces between cells in the parenchyma, and ultimately from the cells themselves, whose inner membrane deflates like a leaky balloon, sometimes tearing in the process. If rain doesn't fall or you wait too long to water, chemical reactions inside cells that depend on water fail to occur. Repeated or extended wilting damages a plant.

The mucigel also provides a hospitable environment for nitrogen-fixing bacteria. Nitrogen is fundamental to life. Without it, neither animals nor plants can build the proteins and DNA they require to function and reproduce. Yet despite the fact that the Earth's atmosphere is 78 percent nitrogen, we animals can't use the nitrogen we take in with every breath. Plants take in air though their stomata but are no better at absorbing atmospheric nitrogen than we are.

The problem is that in the atmosphere, every nitrogen atom is triply bonded with another nitrogen atom. A molecule of two nitrogen atoms (N_2) is that couple on the dance floor, lip-locked and hands deep in each other's back pockets, oblivious to the gyrating crowd around them. The N_2 molecule cannot be chemically incorporated into protein chains and nucleotides. Fortunately for life on the planet, plants *can* incorporate *fixed nitrogen,* which is a molecule composed of one atom of nitrogen bonded with three hydrogen atoms to make ammonia (NH_3). The enormous energy released in lightning can split atmospheric N_2 and allow a single N to bond with three H's. Lightning-fixed nitrogen, however, accounts for less than 10 percent of the total fixed nitrogen on the planet.

The other 90 percent of the planet's fixed nitrogen was

produced solely by nitrogen-fixing bacteria.* Some of these bacterial species live independently in the soil; others live sym-biotically in nodules on the roots of leguminous plants like clover, peas, and soybeans. All the bacteria employ enzymes to slowly transform the gaseous nitrogen in pockets of air in the soil into ammonia and then into nitrate, a form of fixed nitrogen that plants can take up. To get our nitrogen, we must either eat plants, getting our nitrogen secondhand, or eat other animals that have eaten plants, snagging it third-hand.

It takes an extensive root system with trillions of bacte-ria supplying billions of root hairs to gather sufficient nitro-gen, as well as water and other minerals, to support even a small plant. An average potato plant's roots, for example, at the end of a growing season spread five feet in diameter and as much as three feet in depth, and explore about sixty cubic feet—twelve to fifteen wheelbarrows—of soil. Actually, the potato plant is small potatoes when it comes to root systems. The roots of a single bean bush extend into more than two hundred cubic feet of soil, or a cylinder of soil eight feet in diameter and four feet deep. When it comes to tree roots, the numbers are even more startling. The area they occupy underground can be five times greater than the area of the foliage. Contrary to popular wisdom, a tree's roots often

---

* In the early twentieth century, Germans Carl Bosch and Fritz Haber invented an industrial method of converting atmospheric nitrogen to ammonia. The rate at which nitrogen-fixing bacteria can convert nitrogen to N_2 would have put a limit on the planet's human population at about four billion people. Now, synthetic fertilizer made with the Haber-Bosch process provides food for nearly half the world's current population of seven billion. Of course, all the additional terrestrial nitrogen and the increased human population and the concomitant increase in farm animal populations have had significant implications for the environment.

extend far beyond the "drip line," the circle described by a plant's outermost leaves. Roots can extend from the trunk as far as the tree is tall.

About 90 percent of a tree's roots grow in the top eighteen inches of soil, and a good portion of those are in the top few inches. Here is the best grazing for roots. In a forest, many of the tiniest roots actually grow upward, a bit above the soil surface and into a zone of *duff,* which is made of thatch, leaves, and other decomposing plant material. Trees that grow to towering proportions in the forest rarely grow so large or live as long in suburban lawns, in large part because we assiduously remove the unattractive duff, and compact and pave over the topsoil. When it comes to topsoil, "take away the top two inches," writes Robert Kourik in *Roots Demystified,* " . . . and you have a horticultural disaster in the making." We make matters worse by growing turfgrass under and around our trees, so tree roots have to compete with the roots of grass for water and nutrients.

In our neighborhood, the subsoil has a high clay content. Clay is one of the three inorganic components of soil, along with sand and silt. I had thought that clay was a bad thing to have in soil, but the best soil is about 20 percent clay. (Ideally, the remainder would be equally divided between sand and silt.) Of the three components, clay has by far the smallest particles. If a grain of sand were the size of a potato, a particle of clay would be the size of a pinhead. Collectively, the tiny particles of clay have a vast surface area: A baseball-sized hunk of clay rolled out to a sheet a single particle in depth would cover more than an acre. Because each particle carries a slight negative charge and minerals mixed in groundwater have a slight positive charge, the clay particles help keep water and essential inorganic molecules in contact with roots.

When the Carters' builder pushed clayey subsoil on top of the hickory's lateral roots, the clay became a problem. It meant too much water surrounded the roots, filling up the tiny air pockets among the soil particles. The root hairs lost some of their access to oxygen, which they need to burn the carbohydrates stored in roots to release energy. While water passes into the xylem by osmosis, it takes energy to move minerals inside because their molecules are too large to slip through roots' cell membranes. Without oxygen to burn carbohydrates, the hickory was like a starving man too weak to lift the food to his mouth. At the same time, anaerobic bacteria and fungi that thrive where oxygen levels are low probably made themselves at home and settled in to dine on the tree's tender root tips. Gradually, the number of root hairs declined, and the tree had contact with fewer nutrients, including nitrogen.

In addition, as root cells burn carbohydrates, they release carbon dioxide, the same way wood burning in the fireplace or gas combusting in an engine releases carbon dioxide. When the hickory's roots lived in airy topsoil and duff, the carbon dioxide quickly dissipated into the atmosphere. But under three feet of dense soil, the waste carbon dioxide would have accumulated. When a person drowns, he dies not only from the lack of oxygen but from an excess of poisonous carbon dioxide. Likewise, while the tree was drowning, losing its root hairs and weakening from a lack of nutrients, it was also suffocating.

Why didn't the tree die soon after the yard was leveled?

There's no way to know for sure, but most likely the soil was more friable at first and gradually compacted over time. As the soil grew denser, the number of root tips gradually decreased until finally the tree lacked enough nutrients to produce healthy foliage: hence, the two seasons of dwarfed leaves. Last spring, with an inadequate crop of leaves to make new sugars, the tree was finished. The hickory's demise seemed sudden to us, but it had been dying from the day the bulldozer first covered its roots.*

---

* The tree well was also too small. Tree wells *can* save a tree when the grade needs to be raised more than eighteen inches, but only if you provide good drainage and make sure the roots have access to air. According to the West Virginia Extension Service, well walls should extend at least three feet in all directions beyond the trunk's circumference. Before adding topsoil, spread a one- to two-foot layer of rock and gravel over the entire root system and make sure excess water can drain beyond the root system.

ten

Our Fine Fungal Friends

In the early days of my fascination with citrus tree varieties, I came across the strangest one I had ever seen, in the
online catalog of a California nursery. The foliage of *Citrus medica* var. *sarcodactylis* looks perfectly ordinary, but its
fruit is another matter. It looks like the offspring of a large
lemon and a small octopus. The result is a yellow "hand"
with long, pointed fingers, giving rise to the species' popular
name, Buddha's Hand. California growers are permitted to
send citrus trees to non-citrus-growing states, so I ordered a
three-year-old specimen.

The fruit of a Buddha's Hand is bright yellow when ripe.

Citrus trees of that age are grown in five-gallon pots that hold about twenty pounds of moist soil. To reduce the shipping cost, the nursery washes the soil off the trees' roots and encloses them in a plastic bag packed with damp sawdust. The directions that accompanied my small sapling advised me not to use ordinary yard soil or bagged topsoil to repot the tree. These would be too dense and retain too much water, possibly waterlogging the roots. A lightweight potting mix without wetting agents was called for, and I bought a bag from the garden center. I had never dealt with a bare-rooted plant before, but managed to arrange the roots as instructed. After the tree had settled in for a few weeks, I applied the recommended six-month, slow-release fertilizer.

All went well for several months, but gradually the tree began to look a bit tired. The leaves lost their luster, and although it was by then early spring, no flower buds appeared, much less any weird yellow hands. I added more fertilizer, but if anything, the little tree looked worse. I called

Edie for help. The problem, she suggested, was in the soil, or rather what wasn't.

The soil I had used, like many products labeled potting mix, was actually a soil-less mix of composted bark, coir (the fiber of the coconut husk), and the mineral perlite. Such mixes are designed to drain well while retaining an ideal level of moisture. Because the ingredients had been sterilized with heat, my soil-less mix didn't contain soil-borne pathogens, insects or their eggs or larvae, or viable seeds that might sprout and compete with my tree for nutrients. It also didn't have fungi. And that, Edie said, was unfortunate.

You would think all those billions of root hairs would be sufficient to get the job of plant nutrition done. But despite their mind-boggling numbers, they typically contact only about 1 percent of particles in the volume of soil they occupy. That means that a plant has access to only 1 percent of the potential nutrients nearby. Fortunately, roots have developed another strategy for retrieving those nutrients: contracting out.

The soil surrounding roots (the *rhizosphere)* is the most life-packed, biodiverse ecological niche on the planet. Under a microscope, it seethes with microbes, protozoa, and fungi, as well as nematodes, mites, insects, worms, and other creepers and crawlers. It's a zoo down there. Among the estimated 1.5 million different species of fungus trying to grind out a living amid the hungry masses is a group of fungal species called *mycorrhizae* ("my-koh-RYE-zee"). Mycorrhizae are single-celled creatures a thousand times narrower than human hair. Under a microscope, a single fungus looks like a white thread. In the soil, it insinuates one end of itself into a root hair and extends its other end into the soil, where it branches like a microscopic shrub. Lacy, intersecting networks of mycorrhizae, called *mycelia* ("my-SEEL-ee-ah"), are possible to see with the naked eye. Until the late nineteenth century, however, few people remarked on the whitish fuzz they occasionally noticed on tree roots. Those who did assumed the stuff was a harmful parasite or a sign of decay.

In 1885, the Prussian government became interested in truffles, the delicately flavored, highly valuable fungi so cherished by gastronomes. Truffles, round or oblong, grow on the underground roots of certain trees. Specially trained pigs and dogs sniff them out so their owners can carefully excavate them. France had long been a major source of truffles, and by the mid-1800s, some farmers in the southeastern part of the country had even figured out, if not exactly how to cultivate them, then at least how to encourage them to grow. (The trick was to have alkaline limestone soil, the exact combination of hot and dry weather, and to plant acorns taken from the soil near truffle-producing oak trees.) Only one species of truffle grew in Prussian soil, and it had never been the source of revenue that French species were. So, the Prussian government commissioned a well-regarded botanist and professor at the University of Leipzig, Albert Bernhard Frank, to see how Prussia could remedy this exasperating state of affairs. Frank's task was to identify which Prussian trees and soils were best for growing truffles.

Regrettably for the Prussian economy, Frank made no headway in uncovering edible truffles. However, as he investigated the subject, he was surprised to discover that certain thready fungi grew on the roots of almost all the tree species he dug up. He became fascinated with the mycorrhizae (a word he coined meaning "fungus of roots") and realized that they are not only not harmful, but exceedingly helpful—in fact, essential—to their hosts. Not killers or even freeloaders, they are industrious miners that exude enzymes that pry apart organic and inorganic compounds and liberate chemical elements in the soil, especially phosphorous. Phosphorous is an ingredient in DNA. It is also a part of adenosine triphosphate, or ATP, the organic molecule that is the universal store of quick energy that plants (and animals) use to fuel almost every cellular process. After mycorrhizae emancipate elements from the soil, they pass them through their bodies and into a root hair. The threads also act like delicate straws, allowing root hairs to sip from reservoirs of water too minute for even root hairs to access.

Recently, scientists in the United Kingdom have demonstrated that the mycelia also protect plants from predators. When a plant is assaulted by aphids, it produces chemicals that repel them, as well as attract parasitic wasps that attack its attackers. (The wasps lay their eggs in living aphids, which are then consumed from the inside out by emerging larvae.) If the plant is connected via mycelia to other plants, those plants receive a chemical signal from their besieged comrade, and mount their own chemical defense before the first aphid arrives. Nearby plants that are not "networked" do not gear up for an attack. (The aphids, by the way, may counter by producing winged rather than crawling young that can fly off to an unsuspecting target.)

Mycorrhizae don't do all this work for free. Plants compensate their partners by releasing sugars stored in roots. In some cases, they dole out as much as 30 percent of their reserve sugars to their underground partners. The mycorrhizae, nurtured by their hosts, are permanently relieved of the necessity of competing with all those other rhizospheric creatures for organic matter to eat. It is a remarkably productive symbiosis—another term that Frank coined—that can increase roots' ability to absorb water and nutrients by as much as several thousand times. Ninety percent of plants have mycorrhizae, and many can't survive without them.

Paleobotanists now believe that land-based plants owe their very existence to their thready fungal partners. The first members of the plant kingdom to gain a toehold on land more than 450 million years ago were thalloid liverworts. These early liverworts had a flat, nonleafy photosynthetic surface. Instead of roots, they had *rhizoids*—hairlike filaments—that tentatively anchored them to river banks and shorelines. On these wet land surfaces, the rhizoids encountered fungi that had already colonized land some 250 million years earlier. (The fungi literally scraped out a living by exuding a mild acid that dissolves carbon and other elements from rocks.) A little sugar made by photosynthesis inadvertently leaked out of some liverworts' rhizoids; some fungi didn't gather up every bit of the nutrients they wrested from the land's rocky surface; and a marriage of convenience was made. Over millions of years the liverworts' simple rhizoids with their companion fungi grew longer, stronger, and more complex, and evolved into roots. Roots cracked rocks, thereby releasing more mineral nutrients. (Roots wear down rocks five times faster, it is estimated, than weather does alone.) Richer and richer soil supported ever more rooted plants, and the once-barren continents turned green.

One of those species that evolved, hundreds of million of years after mycorrhizae and liverworts first made their bargain, was the *Citrus medica* or citron, one of the oldest extant citrus species, of which my Buddha's Hand is one variety. When the nursery bare-rooted my tree, most if not all the delicate root hairs died, and with them their even more fragile mycorrhizae. New root hairs developed after I transplanted my tree, but because I used a sterile potting mix, there were no mycorrhizae to recolonize the root hairs. Eventually, when I moved the tree outdoors in the spring, fungal spores would likely have landed on the soil, and the symbiosis would have revived. But at Edie's suggestion, I bought a mycorrhizal mix—available in garden stores and via the Internet—and watered it in. Thanks to its fine fungal friends, my Buddha's Hand revived and now regularly produces fruit and its attendant pleasures—zest and pith for cooking and marmalade—throughout the year.

eleven

Arsenic and Young Fronds

About three miles from my home in Maryland is a Washington, D.C., neighborhood known as Spring Valley. Here, multimillion-dollar homes nestle in banks of azaleas and boxwood, shaded by oaks, tulip poplars, and other stately shade trees. Adjacent to the neighborhood is American University, an institution with some twelve thousand undergraduate and graduate students. The manicured tranquility of the neighborhood belies its past. In 1917, while American troops were rushing to the battlefields of Europe, the newly established and financially precarious university contracted its scientists and its undeveloped land to the U.S. Army's Chemical Warfare Service. In order to simulate the western front, the army dug trenches, staked dogs and goats in them, and exploded shells filled with forty-eight poison gases over and in the trenches. The idea was to test the toxicity of the chemicals and to figure out how to defend against them. When the war ended the next year, the army abruptly shut down its operations. It burned seven structures—which, according to newspaper reports at the time, produced a "suffocating" cloud of smoke—and buried leftover shells, drums of chemicals, and contaminated laboratory equipment in pits dug at the outer reaches of the university's acreage. Then, everyone forgot about AU's year of living dangerously.

Decades later, the university sold much of that acreage to real estate developers, and the Spring Valley neighborhood sprang up. In January 1993, a backhoe operator digging a trench for a new sewer line in one of the last undeveloped parcels uncovered four unexploded 75mm artillery shells. The Army Corps of Engineers investigated and discovered a total of 141 buried munitions. The Corps removed tons of soil from eleven sites, and two years later pronounced the area clean. In June 1996, however, workers planting a tree on the grounds of the university president's home were overcome by odors and suffered severe eye irritation. Soil testing at the site revealed hazardous levels of arsenic, a poison that causes cancer and birth defects. The arsenic was undoubtedly a by-product of buried chemical weapons. Testing on the nearby property of the South Korean ambassador's residence revealed arsenic levels that were as much as fifty times the twenty parts per million (ppm) that the U.S. Environmental Protection Agency considers acceptable. After excavation at another residence unearthed 380 shells and several fifty-five gallon drums, most containing mustard and blister gas, the Corps agreed to test every property in Spring Valley. One hundred and thirty-nine lots, as well as AU's Child Development Center playground, were found to have excessive arsenic levels. Much of the ensuing cleanup involved digging up and carting away the contaminated soil, a process that destroyed well-established plantings, damaged the root systems of cherished trees, and left mud wallows in front yards.

In 2004, a Virginia-based company called Edenspace won a contract with the Corps to try a new, experimental approach. It planted the contaminated areas with *Pteris vittata*, commonly known as brake fern. This quite ordinary-looking fern has one extraordinary quality: Its roots pull arsenic out of soil. The arsenic then travels up the xylem to be concentrated in the fronds at levels up to a hundred times greater than in the soil. (The fronds are then harvested and disposed of safely.) Edenspace planted 2,800 brake ferns the first year and more than 20,000 in subsequent years. In 2006, the EPA determined that thirty-five sites at fourteen residences planted in ferns had been cleaned. *Phytoremediation,* the power of plants to decontaminate soil and water, had worked, and the cleanup cost was a small fraction of removing the soil. As a bonus, instead of wallows, the residents had fern gardens.

But why would brake fern roots snarf up arsenic? How could the brake ferns survive laced with so much of a metal generally as poisonous to plants as it is to animals? One of the world's leading experts on the uptake of metals from soil and a founding father of phytoremediation is Dr. Rufus Chaney, senior research agronomist at the U.S. Department of Agriculture. I go looking for him at the redbrick campus of USDA's Agricultural Research Center in Beltsville, Maryland.

The scientist I find in the basement of Building 7 is a ruddy, gray-haired, crew-cut bear of a man who waves me into an office that seems two sizes too small for its occupant. It is lined floor to ceiling and wall to wall with file cabinets and shelves. The shelves are crammed with stacks of papers. Rufus, as he introduces himself, has been working at USDA for forty-three years, and his published work fills a number of those shelves. He has 428 papers and 266 abstracts to his name, and a dozen more papers in the works. In addition to his own research and writing, he lectures frequently, and has directed the theses of three dozen Ph.D. candidates. He could have retired a decade ago, but the work he loves is as yet unfinished.

Nickel has been the leitmotif of Rufus's life, played with variations—mournful, exciting, hopeful, frustrating—for more than half a century. His first encounter with the metal was as a teenager, growing up on a corn and soybean farm in the flatlands of northern Ohio in the 1950s. One year his father applied a liquid fertilizer to their soybean crop, and to his horror, the plants shriveled and died. It turned out that the fertilizer had been delivered in dirty drums contaminated with the waste of a nickel-plating operation.

"At that point," Rufus told me, "my father felt he could gain more by suing the fertilizer company than farming. Unfortunately, that was a very slow process, and he lost the farm to the banks and the lawyers before he could collect. I'd been planning to work on the farm so, of course, it had a huge effect on me: I had to find a different job. It's ironic, but that experience gave me a career."

As an undergraduate, he studied chemistry and wrote his senior paper on nickel in tobacco smoke. As a graduate student at Purdue, he worked on nickel uptake in soybeans, and how it inhibits the uptake of iron, to the detriment of the crop's health. Postdoctoral work at USDA led to a full-time position in 1971, and he was assigned to work on the problem of heavy metals in sewage. At the time, cities were growing rapidly and septic systems that had been built in the previous century were becoming overburdened. Municipalities were replacing them with sewage and wastewater treatment facilities at a rapid pace. But what were the facilities going to do with all the thick residue called sludge (and now, more scientifically but less vividly, referred to as "biosolids") that emerged at the end of the water-cleansing process? Because biosolids are basically human manure, and manure is rich in essential fertilizers like nitrogen and phosphorous, it seemed like a good idea to plow it into farmland.

And it was, in theory. But nitrogen and phosphorous were not the only elements that made their way into the sewers and through the treatment process. Some sludges also contained heavy metals like zinc, cadmium, cobalt, nickel, and others that could kill crops and might, if they entered the food chain, harm people. Rufus was asked to investigate the toxicity of heavy metals to particular crops, whether they entered the food chain, and if there was a way to prevent their uptake into plants.

In 1977, Rufus read a paper on a remarkable quality of *Alyssum bertolonii*, a low-growing plant (unrelated to what gardeners know as sweet alyssum) with masses of small, four-petaled yellow flowers. The paper reported that roots of *Alyssum* took up large amounts of nickel that then accumulated in the plants' tissues at levels that should have been toxic. The plants were found growing on "serpentine barrens," that is, areas where it was generally assumed that the naturally occurring high levels of nickel, cobalt, and chromium (and perhaps low levels of calcium and phosphate) made the soil infertile. Somehow *Alyssum* not only tolerated the conditions, but thrived, never mind that its leaves contained nickel at levels a thousand times greater than average. The paper's authors concluded that "the reason for this preferential accumulation of nickel . . . should prove to be an interesting problem."

Alyssum bertolonii has bright yellow flowers.

Rufus was indeed intrigued. Why would plants have evolved a mechanism to hyperaccumulate a toxin, and how did they do so safely? Serpentine barrens are not common, but there happened to be one not far from his office, and he planted some seeds and began to experiment. In 1980, the EPA became interested in his work, but then the administration changed and the project was killed the next year. Nonetheless, he kept working on unraveling the biochemistry and agronomy of what are known as "hyperaccumulators."

To understand how hyperaccumulators work, we have to drill into roots. Taking up certain chemical elements from the soil is one of the prime functions of roots. Today, the consensus is that seventeen of the ninety-two naturally occurring elements on Earth are essential for all plants. Three of those elements—carbon, hydrogen, and oxygen—plants take from air and water. Of the fourteen soil-based nutrients, six—nitrogen, phosphorous, potassium, calcium, sulfur, and magnesium—are *macronutrients*, which simply means that they are present in plant tissues in large quantity. (Large is a relative term here: Together, the weight of all the soil-derived elements is only about 1.5 percent of a plant's total weight.) The other eight—copper, iron, manganese, nickel, zinc, boron, chlorine, and molybdenum—are *micronutrients*. Some of these are integrated into the substance of a plant's leaves, stems, roots, and flowers. Others are important to processes like photosynthesis and transporting carbohydrates in and out of cells. A trace amount of nickel, for example, is necessary to make an enzyme that breaks down urea, which would otherwise accumulate and damage leaves. An insufficiency of any essential element means the plant grows to less—in some instances, far less—than its genetic potential. (Therefore, when you shop for houseplant fertilizers, look for ones that contain all essential macro- and micro-nutrients.)

In order for a plant to take up any element in the soil,

that element has to be soluble in water. When a water-mineral solution encounters the epidermal cells of a root tip, it flows either directly through them or through spaces between them. Once past the epidermis, it continues through the loosely packed parenchyma cells toward the xylem at the interior of the root. But before it gets to the xylem, the solution encounters the endodermal tissue, a layer of tightly packed cells. Here the going gets tough, at least for the minerals. Any spaces between these cells are tightly caulked by a fatty, waterproof material called *suberin* that blocks fluid from entering. This means that minerals in the water must pass through the membrane of the endodermal cell. While water can pass through freely, larger molecules are stopped. They can only enter via protein-constructed *membrane transporters,* which are something like automated revolving doors. These are not like the revolving doors at airports, however, able to accommodate everyone, including a traveler pushing a luggage cart. Membrane transporters generally fit only one particular element, and the plant creates them or dismantles them in accordance with its need for that particular nutrient. Because plants must spend energy to create and operate transporters, they have evolved to admit only those chemical elements that contribute to the plant's survival.

(The discovery of this dual system for taking up inorganic nutrients—effortless osmosis of some molecules and the selective uptake of others—answered a question that had troubled scientists since the ancient Greek era: Do roots actively choose nutrients from the soil or only passively receive them? The answer, we now know, is both.)

So, why and how does *Alyssum* take up nickel in such

quantities? The first thing Rufus discovered is that nickel is hyperaccumulated only when the soil is acidic, as in the case of serpentine soil. Acidic conditions alter the chemistry of the revolving door for iron so that nickel can slip inside, too. But what about toxicity? It turns out that *Alyssum* has evolved an ability to move the nickel out of its cells and isolate it in its leaves in fluid-filled cells called *vacuoles*. Better yet, once sequestered in the vacuoles, the nickel becomes a homemade pesticide. Nibbling bugs that get a mouthful of nickel-laced leaf either die or look elsewhere for dinner. High levels of nickel actually give hyperaccumulators a competitive advantage. The insecticide is so effective that some nickel hyperaccumulators (and there are nearly four hundred such species in several genera) have lost other disease-resistance mechanisms and cannot survive in soils that have only average amounts of nickel.

It seemed to Rufus that *Alyssum* might be more than an interesting curiosity of botany. It might be a valuable tool in remediating pollution. "I realized," he said, "we might be able to convert this hyperaccumulating weed—because it's a weed until you say it's a useful plant—into something that could decontaminate soils or mine nickel from soils that otherwise aren't any good for agriculture. That was very exciting.

"It became a case for classic agronomy. We found out the fertility needs of our weed, worked out its genetics, and bred to improve it. Since most of the nickel is in the leaves, we selected for those individuals that hang on to their leaves late in the season. We also worked on increasing its yield of foliage and making it taller so standard farm equipment could harvest it. Then we field-tested our cultivars. Not only was it a success; we had an immediate use for the results. In

Georgia, pecan trees die because of nickel deficiency. So, we made an extract of the leaves and used it in a foliar spray, and pecan farmers have found it's a wonderful cure. You can also just spread ground-up *Alyssum* on the soil. You only need to apply a couple of grams of nickel per hectare as a spray, and if you do, you solve the deficiency problem for years. What's more, we can grow this nickel more cheaply than a farmer can buy nickel sulfate."

Rufus leaned forward in his government-issue desk chair. "Then there's the mining opportunity. You grow the crop on nickel-rich soil, then harvest the plants just after they flower, but before they set seed. You cut your crop like hay, bale it like hay, and take it to the biomass energy facility. Then you put the ash into a smelter, and eighteen minutes later"—he leaned back in his chair, and smiled—"you pour out liquid nickel. It's the richest, best nickel ore anyone has ever found."

Every year the United States imports about 125,000 tons of nickel at a cost of about $2.5 billion for use in stainless steel and other alloys, in batteries and electronics, and in construction. *Alyssum* can produce about 350 pounds of nickel per acre of serpentine soil, which means, Rufus says, that U.S. farmers could grow about half the country's annual needs. Since it costs about one hundred dollars an acre to grow *Alyssum,* farming nickel on otherwise useless land could be an excellent business proposition.

So why haven't I heard of nickel farming?

That, Rufus said, is the big mystery. In 1998, USDA, the University of Maryland, and several scientists, including Rufus, were awarded a patent for nickel phytoextraction using these *Alyssum* cultivars. The patent holders promptly licensed the technology to Viridian Resources LLC, a small Houston-based company, with the understanding that it would commercialize the process. Viridian ran some trials at a smelting facility owned by Vale (formerly the International Nickel Company, or Inco) in Ontario that had nickel-polluted acreage. Vale was pleased with the results, but didn't want to move forward until it settled some outstanding lawsuits related to the pollution, which had occurred many years earlier.

"Viridian got five million dollars from Inco to further develop the application at their locations overseas—Indonesia, Canada, and Britain—and they were asked in Brazil to go full speed ahead. But nothing happened."

"Why not?" I asked.

"God only knows," Rufus said. "I'm infuriated. Just think about what this could mean in poor countries like Zimbabwe where they have lots of serpentine soils. They could take otherwise useless acreage that can't grow food and phyto-mine nickel out of it for a profit. Plus, they could then actually farm that land."

Rufus explained that serpentine soils not only have too much nickel, they also have too much magnesium and not enough calcium and phosphorous. Although many plants can tolerate soils with low calcium and low phosphorous, very few can do it if there is also high magnesium. It's another revolving-door problem: Magnesium elbows its way into the calcium membrane transporter and prevents the passage of calcium. Without sufficient calcium, cells can't be created or divide. The few plants that are able to grow on serpentine barrens have evolved transporters that are either particularly efficient at passing through calcium or are less likely to admit magnesium. They are also more efficient at taking up phosphorous.

"So," Rufus continued, "if Zimbabwean farmers mined nickel with *Alyssum,* it would make economic sense for them to add phosphorous and calcium to the soil because doing so would maximize the *Alyssum* and, therefore, the nickel yield. Then, after they've extracted the nickel, the serpentine soil has been fully enhanced and you can grow whatever you want afterward.

"In Indonesia," he added, "people are thinking that you could use it on mining sites. When they mine nickel there, they strip off three to ten meters of topsoil and subsoil, cart it away, do their mining, and then bring the soil back to 're-green' the site. A lot of that soil has enough nickel in it to be valuable. You could spread fertilizer on it, grow *Alyssum,* and then finish off by farming it. Everyone benefits. But Viridian has been doing absolutely nothing with the patent, so none of that is happening."

Later, I tried to reach Viridian, but the CEO never returned my calls or emails. An Internet search revealed a story, though.

In the fall of 1998, Viridian leased some fifty acres from the airport authority in Josephine County, Oregon, and under an agreement with the County Board of Commissioners, planted the fields with *Alyssum* cultivars. In 2002, the company reported that the leaves were accumulating large amounts of nickel, as expected.

Then, in 2005, for reasons unknown, Viridian failed to harvest on time, and mowed the fields *after* the flowers had not only bloomed but gone to seed. The company also left bales of harvested plants, which were full of seeds, piled in the fields. By 2008, *Alyssum* was found at sites far beyond the airport, including on protected properties of the U.S. Forest Service and the Nature Conservancy. The County Board ended the lease with Viridian and demanded the company spray herbicide on the *Alyssum*. The spraying was only partially successful, and the next year the Oregon Department of Agriculture (ODA) dispatched workers to attack the plants again, pulling them out by hand when necessary. The ODA classified *Alyssum murale* and *Alyssum corsicum* as Class A noxious weeds, noting that these European natives are so well suited to serpentine conditions that they threaten to outcompete and overrun unique indigenous flora, including endangered native species that evolved to live only on the serpentine barrens. The cultivars' special vigor, unusual size, and their strong, perennial root systems make them exceedingly difficult to eradicate, although the Forest Service, the ODA, and a corps of local volunteers are still working hard to do so.

Rufus later acknowledged that the *Alyssum* has become a problem in Oregon, but he blames Viridian for failing to cultivate it according to the agreed protocol. If harvested before or even while in flower, he pointed out, the plant has no ability to spread. This is true, but I find it easy to imagine that on a large-scale nickel farm an occasional plant might go to seed prematurely. It could be that the question of phyto-mining nickel will be like other difficult choices we make between economic value on one hand and sensitive ecosystems or rare species on the other. Sometimes the choices are not very palatable. Is it better to employ roots to gently search the soil for nickel even if it also means risking a special habitat? Or is strip-mining the earth elsewhere better? Maybe a choice won't be necessary. The patents that Viridian now licenses will expire in 2015. Then Rufus and others will try to breed a sterile cultivar, one that doesn't produce viable seed. Although Rufus will probably never see any royalties from his years of research, he does hope to see *Alyssum* put to work in remediating nickel-polluted soils.

Although Rufus has not been directly involved with the arsenic-gathering brake ferns, he was instrumental in their story. In the late 1990s, Dr. Lena Q. Ma, an expert in geobiochemistry at the University of Florida, was investigating arsenic contamination of Florida soils and groundwater. At Rufus's suggestion, she searched for species that flourished on the inhospitable grounds of chemical companies, wood treatment facilities, cattle dip stations, incinerators, dumps, and other arsenic-polluted sites. At a facility in central Florida that treated lumber with chromated copper arsenate, a substance once used (but now banned) to make wood resistant to termites, she found the brake fern growing in lacy profusion. Investigating, she found that the plants were sequestering arsenic in the vacuoles of its fronds. The arsenic gained entry to the roots through the revolving door of the phosphorous transporter. Like nickel in *Alyssum,* the arsenic benefits the fern by acting as a potent insecticide. It was Ma's discoveries, inspired by Rufus, that enabled Edenspace to, beautifully, clean up Spring Valley.

twelve

The Once and Future Wheat

The average midwestern farmer tills his fields at least once a season. In the spring, his plow slices through the topsoil, turning the top a few inches to the side to create furrows. Plowing uproots the remains of last year's crop, breaks up fungus-carrying leaves, destroys emerging weeds, and exposes the soil to air, which releases a flush of nutrients. A few days later, after the soil has dried, the farmer may return with a harrow, a piece of equipment with multiple discs or tines that breaks up dirt clods and further aerates the surface of the field. After harrowing, he often disturbs the soil once more, running a cultivator across the land to incise a pattern appropriate for the seeds to be sown.

There are hazards to all this soil disturbance. Mycorrhizal networks are destroyed. Rich topsoil, loosened and sitting exposed on the ground, dries out and is liable to blow away in the wind. On sloping land, rain washes the loose particles downhill. Such soil disturbances and the inevitable erosion, it has been convincingly argued, led to the downfall of the ancient Mesopotamian, Greek, Roman, Mayan, and Incan empires. In nineteenth-century America, farmers plowed under prairie grass in a hundred-million-acre area in the Oklahoma and Texas panhandles and nearby Kansas, New Mexico, and Colorado counties, planting wheat and other crops in its place. From 1934 to 1937, the region was hit by sustained drought, few crops grew, and high winds scooped up 75 percent of the region's topsoil and airmailed it to the Atlantic Ocean. Around the world today, according to experts at the University of California, roughly seventy-five billion tons of topsoil blows away or washes into oceans, lakes, and rivers every year. About seven billion tons comes from the United States. It takes about 250 years for weathering from rocks and organic decay to create a single inch of soil. According to a recent Cornell University study, farming is eroding topsoil in the United States ten times faster (and in China and India, thirty to forty times faster) than nature creates it. In a world that will probably need to feed an additional three billion people by the end of the century, erosion is a global crisis.

The fundamental problem is that modern farming is based on species whose life cycle is complete in one growing season. Most of our edible species—grains like wheat, rice, and corn; legumes like soybeans and alfalfa; and oilseed crops like canola (a rapeseed cultivar) and sunflowers—are annuals. Annual crops cover about 80 percent of the world's farmland and provide 80 percent of the world's calories. This means a global cycle of tilling, seeding, harvesting, and then tilling again the following spring.

This is a most unnatural situation. Without human intervention, most of the world's farmable landscape would be covered by perennials, plants that live for years, and produce new growth primarily from the same, deep root systems. More than 85 percent of North America's native species are perennials. The vaunted fertility of the Great Plains' tallgrass prairie, where topsoil could be as deep as nine feet, was created by perennials. The wild prairie was tremendously productive, supporting a rich diversity of insects, birds, and mammals (including millions of bison) while never requiring herbicides, pesticides, added water, or fertilizers, and all the while banking carbon and mineral nutrients year after year. Now, less than 4 percent of our native prairie remains.

Today, some scientists and plant breeders are making the case that we should re-create a landscape where perennials prevail. If, they argue, we develop deep-rooted perennial versions of our familiar annual crops, we will protect dwindling topsoil, use far less irrigation and fertilizers, save on the fuel required to make nitrogen fertilizer and run farm equipment, and open up marginal lands to agriculture. It is a compelling vision, which is why, in early March, I am driving west from the Kansas City airport to the Land Institute in Salina, Kansas. The Institute, a nonprofit organization founded in 1976 to promote sustainable agriculture, focuses on research and breeding to develop perennial versions of our annual grain and oilseed crops.

I meet my host, Dr. David Van Tassel, in his office in the Institute's research building. David is a rugged man in his early forties wearing a flannel shirt and sporting a stubble of reddish whiskers that reminds me of the stubble in the desolate fields outside. When I rue the fact that I am here in late winter when there is little to see outdoors, he assures me this is the best time for him to entertain a visitor. Right now, he is limited to running experiments in the Institute's modest greenhouse. In a few months, he will be busy outside cultivating, monitoring, and measuring thousands of field-grown sunflowers and bundleflowers, which are the focus of his research.

David proposes we go downstairs to start our tour of the greenhouse. But as we approach the stairs, I am stopped short by a color photograph that spans the height of the stairwell we are about to descend. Pictured life-sized and side by side are the root systems of two plants. The roots of the one on the left are six inches wide at soil level and taper down to a point at about three feet. The roots of the plant on the right form a nearly solid column eighteen inches across and nearly nine feet long. The plant on the left, David tells me, is winter wheat (*Thinopyrum hibernum*), the annual crop that farmers in the region typically sow each September and harvest in May. The plant on the right is intermediate wheatgrass (*Thinopyrum intermedium*), a wild perennial native to the Great Plains. (This is not the wheatgrass sold in health food stores, which is simply sprouts of an annual wheat.) The wheatgrass's roots look like the impenetrable waterfall of beard on R. Crumb's cartoon character Mr. Natural. In comparison, the annual wheat's roots look like the ends of an old man's Fu Manchu.

Once I cease exclaiming about the astonishing difference

between the plants in the photos, David tells me there's more to the matter than appearance.

"You have to remember that annuals' roots only reach that three-foot length shortly before they begin to die. Even more important is that annuals' roots are not in the ground for much of the year. The way we grow wheat around here, for about nine months of the year the ground is either bare or just seeded, and there aren't any roots in there at all."

The wild wheatgrass root system, on the other hand, stays in place all year. Its roots reach far more pervasively through the soil to access more water and nutrients than annuals', enabling them to grow on more marginal land. The perennials also corral more phosphate and nitrate fertilizers, which otherwise easily leach out of the soil and pollute groundwater or cause excessive algal growth in lakes and ponds. In fact, perennials capture twice as much of applied fertilizers as annuals do.

Because perennial roots and their mycorrhizal partners merely slow down or go dormant in winter, the plant is ready at the first hint of spring to send up new stalks. Wheatgrass is up and busy photosynthesizing in full force many weeks before the stalks of a newly sown annual variety appear. It stays green even after its seeds fall, continuing to manufacture carbohydrates. Some of those carbohydrates are stored in the roots, which, like soup kitchens, dole out a certain percentage to feed the multitudes of organisms in the rhizosphere. Those burrowing, digesting, and decomposing creatures create the granular texture of good soil and, at their deaths, build its fertility. That fertility in turn supports plants and a diverse aboveground ecology. And, of course, perennials' roots are active year-round in holding soil in place.

So why, I ask, aren't all our grain crops perennials, especially given that in the wild they are so much more prevalent than annuals? The quick answer, David says, is that annuals tend to channel more of their energy into producing more and bigger seeds—the part of the plant we use for food—than perennials do. A fuller answer awaits us downstairs.

At the door to the greenhouse, I look into a sea of six-foot-high grasses with their green stalks and narrow, blade-like leaves. All the plants are growing in ten-gallon black pots and all the pots are arranged, rim to rim, in shallow metal trays. Right now, the trays have only a little water in them, but in a few hours they will be flooded so the plants can absorb water from their roots. Grow lights hang over the plants, but at noon on this sunny day, they are outshone by natural daylight.

Narrow aisles separate the trays, and we sidle into the wheat section. "In this section," David says as we proceed crabwise, "we're crossing annual winter wheat and wild perennial wheatgrass. Because each seed has its own particular genetic inheritance, any cross between two parents are going to produce a wide range of hybrid individuals. It's a roll of the genetic dice: Some individuals are going to be almost like wheat, others almost like wheatgrass, while most will fall in a bell curve of intermediate traits."

He stops and bends a sturdy stalk toward me that has, at about my eye level, a curving seedhead packed with plump, pale golden brown seeds nearly the size of rice grains. The tip of each seed has a long, stiff hair, called an *awn*. (The awn probably makes it harder for birds to land and snatch the grain.) "This plant," he says, "is very close to the annual species. If this plant were growing outside, this is how it would look in May, right before harvest. You can see how large the seeds are, and that they're all still on the head."

We sidestep a few steps farther down the row as David looks for another plant to show me.

"Now take a look at this one. It's a lot closer to its wheat-grass parent." When he pulls this flower stalk toward me, I see that its seeds are arranged, not in a dense, three-inch crescent, but stretched out along the top twelve or so inches of a stalk. The seeds are obviously smaller, too, more like sesame seeds.

Wheatgrass　　*Winter Wheat*

On the left, seedheads of wild perennial wheatgrass; on the right, seedheads of domesticated annual winter wheat.

"These seeds are a problem because there's less edible germ and more inedible hull in each one. Another problem is that when they mature, they probably won't ripen all at once, and when they do, they'll drop to the ground right away. It's what we call shattering. From the farmer's point of view, this is no good at all. He wants to harvest once and capture all the ripe seeds that one time. The ideal seedhead—again, from a farmer's point of view—is the modern ear of corn. Its seeds always ripen at once and they *never* fall off."

While a habit of shattering is bad for farmers, it's a competitive advantage in the wild. The prairie landscape is dense with perennials; normally, every fertile spot is taken. When the rare vacancy occurs—maybe a tunneling gopher destroys a plant or a herd of bison tramples down a swath of grass—a plant with a shattering seedhead is more likely to have some ripe seeds ready to seize the moment. The smaller size of a perennial's seeds is also the result of evolutionary selection. Perennials divide their energy and carbon between building the roots that are the source for sending up new stalks in the spring and seeds that might populate a bare spot should one happen to open up. So, perennial plants generally produce shattering seedheads with small seeds. Annuals, whose only hope for survival lies in their seeds, produce big seeds packed with nutritious endosperm.

Crossbreeding is one way to try to perennialize an annual. The other way is to select and propagate the most desirable of the wild perennials, then choose the most desirable among those to propagate, select among the second generation, propagate again, and so on. The Institute's scientists are pursuing both methods simultaneously. I was thinking that a little gene splicing could make quick work of the problem. But there is no single gene for perenniality, David tells me, any more than there is a single gene for flight in birds. Perenniality involves a complex suite of attributes. You would have to identify and manipulate untold numbers of genes that control for seed size and ripening timing, root growth, seedcoat thickness, and allocation of carbohydrates between roots, shoots, and seeds. Genes for stalks would have to be manipulated, too, for the sake of a mechanical harvester. Wheatgrass stalks are naturally tall, thin, and variable in height. To be fit for the farm field, they must become shorter, stiffer, and more uniform. Old-fashioned breeding techniques are more likely to do the job.

The seeds of wheat, rye, oats, sorghum, barley, and other grass plants store energy in the form of carbohydrates, carbohydrates we snatch and turn into breakfast cereals, bread, and beer before the plants' embryos and seedlings can use them. The seeds of other plants, like rapeseed and sunflower, fuel their embryos with energy stored primarily as oils. Oilseeds are critical to human health because they contain omega fatty acids that we need for proper metabolism but our bodies cannot manufacture. We have also come to value the oil as a cooking medium: Not only does it impart taste, it transfers heat efficiently. The sunflower (*Helianthus*), which produces an oil particularly low in saturated fat, is the chief focus of David's research, and we wend our way to the far end of the greenhouse to see what he has growing.

I find sunflowers creepy. From time to time, Ted grows one in his postage-stamp vegetable garden. With its man-sized head, large and prickly leaves, and a towering stalk that seems too weak to support its heavy flower, the mature plant looks like a freak of nature. (The flower is not one bloom, but rather is made of hundreds of its tiny "disk florets" encircled at the perimeter by far larger, single-petaled "ray flowers.") At the height of summer, I admit, a sunflower in bloom is a striking sight, the bright yellow petals rimming a burnished, golden disk. By late summer, though, the florets have turned to black seeds and the ray flowers are withered. From our back windows, the flower looks like a dark, protuberant eye rimmed by lashes staring blindly at the house all day. It disturbs me to see birds fly in to peck at its undefended oculus.

I am gratified, therefore, to learn from David that the annual garden sunflower *is* a freak, not of nature but of man. In the wild, most sunflower species look more like daisies and other charming members of the aster family. The flower's disk is a modest inch, rather than twelve inches, in diameter. If a wild sunflower finds itself in a good patch of sun-drenched soil, it will branch prolifically, becoming a head-high shrub sporting many dozens or even hundreds of small flowers. One well-situated wild sunflower plant can produce tens of thousands of seeds.

On the left, a wild sunflower; on the right, a domesticated sunflower.

Helianthus evolved in the American Southwest, and later spread across the American plains, courtesy of the great herds of bison. Their matted coats snagged the little hairs at the ends of ripe sunflower seeds and carried them away. When the seeds dropped off, they fell on ground that their chauffeurs had tilled with their hooves as they passed through the high grasses. In a matter of weeks, stems and leaves would rise out of the trampled earth. It was as if the bison left a trail of sunflowers, like bubbles in the wake of a boat. Some ten thousand years ago, Native American tribes began migrating across those same lands, along the bison trails, and gathered the ripe seedheads. The seeds could be hulled and ground into flour or mixed with berries, meat, and fat into a sort of energy bar called pemmican. The hulls, brewed in hot water, made a gorgeous purple dye.

The gatherers inadvertently began to domesticate *Helianthus* long before anyone deliberately planted it. Picture a midwestern landscape four thousand years ago, lush with wild grasses and wildflowers. A tribe has set up a seasonal camp, and a woman sets off for an afternoon of collecting sunflower seedheads. She has noticed that there are patches of sunflower plants that have somewhat larger heads and seeds, some up to twice the size of others, so she favors those, although there are far fewer of them. Any seedheads that have already dropped a good portion of their seeds or that are still partially in flower, she ignores. She may simply pull the seedheads off or perhaps she cuts them off with a stone blade. Then she drops them in the basket she carries in her hand; when it's full, she empties it into the much larger basket hanging on her back from a strap across her forehead.

Those attractive patches of *Helianthus,* the ones with the big flowers and seeds, are annual species. Annuals represent only 10 percent of sunflower species. They pour their energy into reproduction like there's no tomorrow because, in fact, there is no tomorrow. The only future annuals have is in their offspring. So they rocket their shoots and leaves skyward, make minimal investment in roots, get to reproduction as fast as possible, and channel their energy into lots of large seeds. Naturally, these are the species that draw the eyes of our human seed collector, although she undoubtedly collects perennials' seeds as well. Back at the camp, she harvests the seeds and tosses the (mostly) empty seedheads onto the camp's trash heap.

The next year, a few sunflower plants—"volunteers"— spring up on the trash heap or where a gatherer dropped a few seeds elsewhere in camp. The volunteers are more likely to be annual species, both because the gatherers favored them and because annuals flourish in such open, disturbed ground. The trash heap plants tend to have nonshattering seeds because they descended from parents that were harvested for that attribute. Over time, a convenient patch of annuals, cross-pollinated by insects that have visited their similar neighbors, persists.

In the wild, annual patches are crowded out over time by perennial species. Perennials are slower off the mark in the first year or two, but once established, in the spring they send up new shoots from their roots faster than annual seeds can germinate, and after several seasons shade out the annuals. In the wild, perennials ultimately win the race, which is why 90 percent of *Helianthus* species are perennials.

Why the first hunter-gatherers bothered to deliberately plant sunflower seeds when they were available in the wild is a matter of debate. Did a tribe move to an area where no convenient patches of the annuals grew? Lazy soul that I am, I think someone simply appreciated the convenience of the camp patch, and sowed some saved seeds or expanded the local patch of volunteers. In any case, roughly four thousand years ago, the sunflower was domesticated somewhere in the region of southern Iowa and western Tennessee.

When it came time for planting each spring, the neophyte farmers naturally selected the biggest of their big seeds to sow. In choosing the biggest seeds, they inadvertently chose other plant traits. The biggest seeds often came from individuals whose genes had directed them to grow fewer branches and devote more energy, therefore, into fewer flower heads. The plants with fewer flower heads had more energy to put into producing bigger seeds. By repeatedly choosing to plant the largest seeds of the fall harvest, Native Americans gradually developed a minimally branching plant with foot-wide flower disks that have large, nonshattering seeds.

The question at the Land Institute is whether modern breeders can go back to the beginning and start the process of domestication over. David is crossing wild perennials and domestic annuals in the greenhouse, but he is particularly enthusiastic about domesticating a wild perennial called *Helianthus maximiliani*. The *maximiliani* branches luxuriantly in nature and sports multitudes of small flowers that shatter freely. After only four seasons of selective breeding, David's *maximiliani* have already developed a nonshattering trait. He shows me two plants that also have a single, unbranched stalk and a head about twice the size of a wild *maximiliani* head. He isn't sure, though, that ultimately a domesticated perennial will have a single head. When farmers harvested by hand, a single huge flower made harvesting easier, but today's mechanical harvesters don't care whether they strip one head or a hundred heads off a plant. Besides, there is a downside to the giant heads: They make great perches for marauding birds. In commercial fields along migratory bird routes, farmers can lose up to 40 percent of their sunflower crop.

"I believe," David says, "domestication is going to be easier than people think. People assume that because it took place thousands of years ago that it *took* thousands of years, but we don't actually know how long it took. Now that we know how plants work, about sexual reproduction, about genes, and have sophisticated breeding techniques, we should be able to do it a lot faster than the people who did it accidentally. Now that we have people devoting their whole lives to domesticating these plants, I'm betting it will happen a lot faster."

Still, the breeders here talk in terms of twenty-five to fifty years. So far, the best of the Institute's perennial wheat produces about three hundred pounds of grain per acre, a trifling amount for Kansas farmers who currently harvest more than two thousand pounds per acre. David points out, however, that a perennial wheat crop needn't be quite as productive as the annual it replaces in order to be profitable. Perennial farmers won't have to buy new seed and will spend less on fertilizer, machinery fuel, and irrigation. Still, he and his colleagues have a long way to go. And it is no sure thing. Skeptics believe that creating perennials with big, calorie-rich seeds may be like having your cake and eating it, too, a logical impossibility. In nature, they argue, species make a trade-off between living a long life and producing many offspring: With the same amount of sunshine-generated carbon, species "choose" between investing it in big perennial root systems or big seeds.

It may be, however, that Land Institute scientists manage to shift that carbon allocation, creating a variety that puts just enough energy into roots to survive the next year while channeling substantial energy into creating seeds that are nearly the size of annuals'. Or, there may be carbon savings to be found elsewhere in a plant, say, in stem height. Wild grasses, having evolved in prairies under natural selection, developed tall stems to beat their neighbors to the sun and avoid being shaded. If breeders select for perennial wheat that is shorter, will they be able to breed plants that redirect that carbon into seed production? David acknowledges that he and his colleagues are taking a gamble. But if they succeed, they will not only revolutionize agriculture but also change the future of our planet.

thirteen

Off to the Races

Let's say you start a giant pumpkin in your backyard in the spring. Throughout the summer, you tend your patch assiduously, burying the vines' axils, fortifying the soil, and drenching it with warm water. Now it's the end of the season and your pumpkin has reached her voluptuous maturity, and you've got somewhere between four hundred and twelve hundred pounds of pulchritudinous orange gourd. She's big all right, but probably not big enough to win any prize money at a regional weigh-off. However, if you happen to live near Damariscotta, Maine, a coastal village an hour north of Portland, you're in luck. You have an opportunity for fame, if not fortune. Saw off the top of your pumpkin, scoop out the innards, and get you and your mama down to the harbor to the Giant Pumpkin Regatta.

This is where Ted and I are at eight o'clock on a Sunday morning, a morning as fine as Maine can manufacture in mid-October. The sky is a kindergartner's vision of blue, the temperature is pushing seventy, and the glassy water of the harbor doubles the hulls of colorful skiffs at anchor. We arrived early to find a spot along the riprap where, I was forewarned, a crowd of hundreds would gather. To my right is the town boat ramp, where a shiny green John Deere tractor is grinding its way down to the water's edge with a giant-pumpkin boat strapped to its front forklift. This pumpkin has been painted a pale gray with a grinning mouth full of big white teeth on its front end. On its back end is a perky Styrofoam whale tail. Standing in the water is a member of the town's fire department, a well-fed fellow wearing a red drysuit. When the whale floats free of the forklift, he swims it over to the narrow floating dock. There, its owner, a man in a baseball cap, reclaims it.

The whale has company. To its left is a pink pig with a pink bucket for a snout, googly plywood eyes and upright ears, and a curlicue wire tail. Bobbing gently on the right is a pumpkin peacock with a towering Styrofoam head and arcing tail feathers made of blue, pink, and green swimming pool noodles. The whale's owner jackknifes himself into his craft, takes up a double-bladed kayak paddle, and sets out for a test drive. When he returns, he starts hacking off pieces from the rim of the cockpit. Meanwhile, some official is shouting that everyone who is not a participant in the race has got to get off the "dawk," which, he correctly points out, is sinking under their weight.

By nine thirty, hundreds of people have indeed gathered on the shore, and an unseen announcer asks us for silence.

Hands over hearts, we sing the national anthem. On a yacht at
the far end of the dock an American flag is hoisted by members of a Boy Scout troop. Three Marines in crisp blue pants,
dark jackets, white belts, and white caps stand at attention as
the flag goes up. A minister blesses the pumpkin growers and
carvers, the sailors, the festival organizers, the observers, and
the Damariscotta fire department. Then, we're off to the races.

There are two classes of pumpkin boats in the regatta:
paddle-powered and motor-driven. First up are the paddlers.
I see that the two smallest craft—the pink pig, which is cap-
tained by a tanned young woman wearing a pink ball gown
and an orange beanie, and a pumpkin painted to resemble a
pink-and-white-striped hard candy—ride lower in the water
than the others. There's a bit of jockeying before the start. A
particularly plump pumpkin, one that looks like an orange
washtub, is drifting away from the dock, and its captain can't
seem to paddle in reverse to return to the starting line. Suddenly, despite the washtub's two-yard lead, the race begins.

The whale quickly passes the washtub, which doesn't go
forward well, either. The pig is in hot pursuit. The Maine
Maritime Museum's entry, "Smash," a pumpkin completely
hidden by a two-masted, yellow-hulled brigantine complete
with flapping sails and an American flag, can't get away from
the dock at all. The candy and the peacock are struggling.
(The peacock's majestic neck prevents the paddler from
reaching far enough forward to get a good purchase on the
water.) Seventy-five yards into the race, as the boats round
the buoy to head home, things get messy. It seems there is no
protocol for the turn: Some boats go to the starboard side,
others to port, and then the buoy drifts, and there is a general

mash-up as a result. The captain of the candy, the only one who executes a quick, neat turn, emerges first from the pack and moves instantly from fourth place to first. The peacock, whose head is loose and swings drunkenly at each stroke, falls to the rear. The candy sprints for home, but is beat out by the whale by a length. The crowd cheers, and I can hardly aim my camera for laughing.

In the second heat, the pig starts taking on water, which, someone with binoculars tells me, is entering through the base of its tail. A little girl behind me gasps, "The princess is swimming!" and indeed she is, dog-paddling away from the swamped pig. Not to worry, though. Two frogmen escort her to shore and push the pig back to the boat ramp. I am soon distracted from this drama, though, by events on, or rather under, the brigantine. This time, the ship did leave the dock, powered with two oars by a young woman dressed in period costume, but she is still on the outgoing leg while the others are on the incoming. Despite the captain's ardent rowing, her ship is faltering and seems to be settling in the water. Suddenly, an orange blob arises from the deep, next to the ship and just beneath the surface. It is the brigantine's pumpkin, which has parted company with the superstructure. Once again, the intrepid frogmen swim to the rescue. Meanwhile, the peacock—which, mercifully, has been decapitated between heats—is making a strong showing, but the whale wins again.

Next up is the motor-driven-pumpkin race, which I am sorry to report goes off without a hitch. Only the largest of pumpkins, the thousand-pounders, can support the wooden platform that rings the gunwales and bears the weight of an outboard engine. Some of these giants are big enough that their pilots, all of whom are male, can stand up in the pumpkin and steer. Several boats have superstructures that all but obscure the vegetal nature of their craft. There's one that looks like a psychedelic Boston Whaler; another could pass for one of the skiffs anchored in the harbor. One pumpkin is surrounded by what looks like a giant black bat wing. I see that the engines are of varying sizes.

The men throttle up and the boats are off on a course that calls for five laps around a buoy about 150 yards from the dock. Most of the men take on that determined, forward-leaning pose you might see in a Winslow Homer painting of a fisherman plowing through the frothing waves of a Nor'easter. I am rooting, however, for the captain of an unadorned pumpkin that looks like a gigantic, buff-colored cauldron. He is a slight man with a long gray beard and a round belly. Dressed in an orange T-shirt, green suspenders, and a black Greek fisherman's cap, he perches perfectly upright and calm at the helm, as if teleported from a church pew. The cauldron, riding high in the water and with only a Mercury 3.3-horsepower engine, is passed by everyone, and soon is lapped by all the boats, even the underpowered bat wing whose captain is standing and pretending to lash the sides of his craft as if it were a horse. The cauldron bobbles from side to side in the passing wakes, but the sea captain steers straight on, serene and unperturbed.

My favorite eventually arrives dead last and at least two laps behind the others. It is clear to me that horsepower is the decisive factor in the race—the winner has a fifteen-horsepower Evinrude—and I comment to Ted that there should be a rule limiting the size of the engines. He tells me that no right-thinking Mainer would consider such a thing: May the man with the biggest machine win.

After the races, I happen upon the owner of the whale, Peter Geiger. Geiger, who has driven up from Lewiston, Maine, where he is the editor and owner of the *Farmers' Almanac,* was also a victor at last year's regatta, which he won paddling a pumpkin decked out as a NASA space shuttle. One of the keys to success in a pumpkin race, he explains, is getting the right-sized pumpkin and getting it balanced properly. At first, his whale had listed to one side, which is why he cut hunks off one side of the cockpit. Sometimes people take on bags of sand at the last minute for the same reason. A few years ago, he had an 870-pound pumpkin, and the boat, painted as a cow, was so big and unbalanced that every time he got in it, it flipped him over into the water. I shudder at the thought: The water temperature in October is in the mid-fifties.

After the races, Ted and I wander through Damariscotta, one of the loveliest villages in Maine. The Pumpkinfest has taken over the town. Lining the sidewalks are sculpted or painted giants, sixty-five in total. We stop to watch an artist incise an elaborate bas-relief of a coastal marsh scene into the surface of his four-hundred-pounder. The 1,375-pound champion of the local weigh-off, which was held last weekend at the start of the festival, is enthroned in a hay wagon cushioned by straw and surrounded by purple mums. Grown by Elroy Morgan, it is a beauty, waxed to high sheen and exuberantly orange, the very essence of pumpkin.

We have already missed the all-you-can-eat pumpkin

pancake breakfast, the pumpkin parade, the pumpkin derby, and the pumpkin pie eating contest, so we're glad to see that we're still in time for the pumpkin catapult and pumpkin hurling. These events take place a few miles out of town, in a grassy field that slopes down to Salt Bay, which lies as smooth and shiny as a pool of mercury. We park our car along Mill Street and join the amiable flow of couples and families, many with picnic baskets and blankets, dogs on leashes, or children in strollers. It's postcard New England, and we arrive just in time for a catapult launch.

The catapult is a two-story triangular contraption, featuring an impressive array of heavy-duty springs. The springs are stretched taut by cranking back the launching arm, and when the arm is released with a reverberating bang, a pumpkin soars thousands of feet overhead and lands in the bay with a splash you might reasonably expect from a breaching whale. The crowd responds with whoops and applause. I ask a man standing beside me, evidently enjoying the event, if he's ever seen anything like this before. Oh, no, he says, this is his first time; he's only seen it done with bowling balls. Clearly, I've got to get out more often.

Even better is the pumpkin hurling, which involves a makeshift cannon. The barrel of the cannon is about thirty feet long and is paired to what look like two jumbo-sized household water heaters stacked on their sides. These tanks are connected to a well-drilling machine, a sixty-thousand-pound piece of equipment with a roaring air pump that normally would drive a drill bit. Instead, the pump channels air to the tanks. The barrel, breech-loaded with an ordinary field pumpkin, is aimed at a decrepit gray Toyota pickup truck at the lower end of the field overlooking the water. After the pressure has built up in the tanks and they start groaning, a bearded man wearing neon-orange ear protectors, a yellow hard hat, and a T-shirt that reads "Punkin' Chunkin' World Champion" fires the cannon. It goes off with a terrific boom. Instantly, automotive parts go flying. I witness the passenger door take flight, the hood somersault away, and a spray of metal innards explode from the engine compartment. At the bearded man's signal, the children charge down the field to inspect the damage, followed by adults who saunter down the greensward after them. The event is so harmless and so silly, the day so glowing, and the spirit so Rockwellian, I wonder what it would be like if we moved to Damariscotta. If we did, we would increase the town's population to 1,912. Which raises the question, how did such a small town, and one noted for its maritime heritage, come to have a pumpkin festival that draws some twenty thousand visitors to its quaint streets?

The answer, I am told, is Buzz Pinkham and Bill Clark. Buzz is the earnest, clean-cut, blue-eyed owner of Pinkham's Plantation, the local garden center, and Bill is a laconic, dark-haired, and bearded naval engineer employed at the Bath Iron Works. I caught up with them and a few of their pumpkin-boat-building buddies in the late afternoon at the back of Pinkham's. The gist of the story, told by Buzz in an accent that makes only a passing reference to his *r*'s, is that about eight years ago, Bill bought a handful of giant pumpkin seeds and planted several in his backyard. They grew pretty well, and one day he was down at Buzz's picking up some fertilizer when the two of them started talking pumpkins. Bill said it wasn't hard to get little plants going, and it dawned on Buzz that giving away giant pumpkin seedlings might be a good promotion for the garden center.

The first year, he gave away a hundred seedlings. The next year, he prepared two hundred seedlings, which were gone before nine in the morning. Pretty soon, he was giving away many hundreds of seedlings. Busloads of schoolchildren would come in to the store in spring to learn how to care for pumpkin plants, and he was tickled with the opportunity to teach them about the way plants work. He also sold a fair number of Langevin's book on growing giant pumpkins. One day, he and Bill were looking through the book and saw a photo of a pumpkin boat. That tickled the naval engineer, who said he'd make a boat for Buzz, if Buzz would row it. Buzz, who I imagine has never turned down a chance to try something new and potentially fun, was game. (Buzz, it turns out, was the jokester lashing the bat-wing motorboat.) People heard about it and came down to the harbor to watch, and the Damariscotta regatta was born.

Buzz and Bill and a few others got the idea of growing pumpkins for charity. Now, in the spring, kids sign up relatives, neighbors, and businesses to pledge a sum per pound of full-grown pumpkin. Half the take goes to a charity of the grower's choice; the other half goes to the Pumpkinfest committee activities, including the weigh-off prize money. (The winners go home with a portion of a ten thousand dollar pot.) Every year, the number of visitors has grown.

"The thing just sorta caught on," Buzz said. "Take Dan here," and he nodded at a gawky man standing near us. "He just got fourth place in the weigh-off. Now, he used to be a stock car driver. His whole family was into stock cars, and their wives were stock car widows. Every year, the wives'd go over the budget, saying"—Buzz exaggerated his accent—"How much d'ja spend on that stahk cah this year? How much d'ja win? You're not gonna do it again this yeah, are ya?' "

Buzz continued, "And then Dan and a lot of others, they discovered pumpkin growing. The wives thought that was really good. They'd say, 'What? The pumpkin seed's only sixty bucks? Oh, great, why don't you get two or three? You need a load of manure for the gahden? How much is that? Only a hundred bucks? Go right ahead.' This was nothing compared to a new transmission or a new set of tires.

"Of course," he added, "it can get a little out of hand. I mean there's the unfortunate wife who hadn't been sub-jected before to any of her husband's hobbies. Then pump-kin growing comes along, and she thinks it's the worst thing that's happened in the world. Why? Because all of a sudden her husband's in the pumpkin patch all the time, and he's watering and putting out manure and he's on the Internet and he's in the pumpkin chat room and he's getting the soil tested. The house could be falling down around their ears, but that pumpkin patch sure looks good." Everyone laughs knowingly.

When it comes to pumpkins, Buzz says, everyone is always talking about how to get an edge. Two years ago, he began inoculating his soil with mycorrhizae and promptly grew his best pumpkin, a 1,266-pounder. (Sadly, at 1,099 pounds the pumpkin developed a split in its side. Although he fixed it with the wax ring from a toilet installation kit, a patched-up pumpkin was ineligible for a prize.) Still, while a competitive spirit spurs on many people—Mainers will compete on almost anything, including table-saw and belt-sander races—it's a small component of the pumpkin mania in Damariscotta.

"The interesting thing is," Buzz said, "it's a real community builder. Some neighbors can live near each other for years and never talk, but this gives them something to chat about. People who pledge a pumpkin want to see how it's doing through the summer, and the pumpkin sort of develops a fan base. Then everyone turns out at the festival to see how 'their' pumpkin's been decorated."

Giant pumpkin growing involves nearly everyone in Damariscotta. Kids and adults grow them, local artists craft them, businesses support them, boat owners lend their engines for the race, charities arrange pumpkin transport, the Shriners parade, and a volunteer committee of dozens works through the year to organize the festival. Even the local curmudgeons get involved, writing cranky letters to the *Lincoln County News* about the traffic jams during the festival. It's as if the pumpkins' giant root systems twine from yard to yard and stretch from the harbor throughout the town and out to Salt Bay, pulling townspeople and tourists into one productive, if rather goofy, enterprise.

PART III

Leaves

fourteen

New Beginnings

After college, I did not become a poet; I never tried. The thought of starving in a garret, as my father warned me I would, scared me, and there was no way I was returning to Baltimore to scribble in my childhood bedroom. Like many a recent college graduate, I had no idea what I wanted to do for a career. I only knew what I couldn't or didn't want to do. Law was a nonstarter, thanks to an aversion to conflict and a well-demonstrated inability to pay attention to detail. Medicine was impossible: I have always been profoundly grateful that our internal anatomy is neatly packaged out of view inside our skin. Even the sight of my own meandering blue veins makes me slightly queasy. Science was out (see "dexterity, lack of"). Business held no attraction: My father was already counting the days to his retirement. As for journalism, I grew up in the era of Watergate,

Woodward and Bernstein, and investigative reporting, and knew I didn't have that kind of bold nature. At five feet, one inch, and weighing ninety-five pounds, I was congenitally unfit for any job involving heavy lifting. The only thing I could competently do was read and write.

One day in my senior year, I was in the career office, morosely flipping through binders of jobs I wasn't qualified for, when I saw an announcement for an internship at a foreign policy think tank in Washington, D.C. As an American studies major (with a heavy emphasis on American literature), I knew it was a long shot on the face of it. Naturally, the application called for relevant writing samples. I had taken one course in international politics, so I had one term paper to submit. I should have added another from an American history class, but for some reason I saw fit to enclose a book review I'd written for a campus literary magazine. The book in question was Nora Ephron's collection of essays *Crazy Salad,* and I had highlighted the chapter on her interview with Linda Lovelace, star of the hard-core porn film *Deep Throat.* I still have the book, and remember quoting the following exchange: Ephron: "Why do you shave off your pubic hair in the film?" Lovelace: "Well, it's kinda hot in Texas." It seems clear in retrospect that this was not the ideal writing sample, but whoever at the think tank had the task of reading scores of college term papers must have appreciated a dash of comic relief. I always tell people that Nora Ephron got me my first job.

The internship was perfect for me; I did research for a book on the CIA's dealings with the Kurds in Iraq. That job led to other jobs, including ones at a congressional committee on foreign relations and a foreign aid agency, then to a master's degree in international relations, and finally, for fifteen years, a position with a government corporation that facilitates investment in developing countries. I traveled in Latin America, the Middle East, and East Asia, learned how to read a balance sheet, write contracts, focus on details, give speeches, ask uncomfortable questions and listen critically to answers, critique and take criticism, manage a team of employees and contractors, and enjoy the camaraderie of smart and committed colleagues engaged in a meaningful endeavor.

Yet somehow, at forty, slowly and without even a telltale hiss, the air seeped out of the enterprise for me. The script was great, my role was a plum, the pay was good, but my heart was no longer in it. I had to reconsider. I'd already given up travel because I hated to be away from our three daughters even overnight. Now that all the girls were out of their baby years and more engaging by the minute, I wanted to spend more time with them. I also realized that my greatest pleasure at work increasingly came from writing. Whether it was crafting an internal memo or detailing an employee's accomplishments or making the case for board support for backing a business in Sri Lanka, the writing itself had come to matter as much to me as the substance. With Ted's support, I decided to try a freelance writing career.

I was glad to write on any subject for anyone who would pay anything. I wrote articles for newspapers and magazines about the trucking industry, swim instruction, my mother's cooking, the "Help me, I've fallen" devices, portraiture, cochineal dye, and a glass capsule of "liquid skunk" that I bought as a personal safety measure (break it and your assailant flees in disgust) but then lost somewhere in our house. I wrote school newsletters, corporate annual reports, and papers for a congressional watchdog agency. One of the latter I converted into a book for school libraries on the history of the U.S. census, just in time for our decennial enumeration. I was pleased; I was writing. But I began to wonder: Was the process the only goal? What was it I really wanted to write about?

Epiphany came via our daughter Anna and the local elementary school. Every year, the fourth and sixth graders participated in either a science fair or an "Invention Convention." When Anna was in fourth grade, it was the year for the latter, for which every student had a few weeks to invent something. If there were any more explicit instructions or guidance, she never told me and I never found them in her backpack. In any case, Anna was sure about what she wanted to create: a device that would allow kids to read in bed with a flashlight without getting in trouble with their parents. This involved attaching a long string to the handle of the bedroom door, routing it through pulleys on the ceiling, including one over the nightstand, and tying it to the back end of a flashlight. When a parent pushed open the door to check on the child who should be sleeping, the flashlight, which had been hanging above the nightstand to illuminate the book, would dip into a large mug, effectively extinguishing it. Anna needed a scale-model of a bedroom with the device to take to school, so the project required Ted's involvement as he was the only one in the house who could be trusted with a saw.

Daughter and father had a great time together, but I wondered what the point was. There was no connection to her science class or to understanding the process of invention.

Why, for example, was using pulleys (which I suspect was a suggestion from her dad) instead of hooks a good idea? The night of the convention, it was clear that not every child had a workable idea or a dexterous enough adult to make a presentable project. Six years later, our youngest daughter, Alice, participated, and for her the convention was meaningless. A decade later, she can't even remember what she made. But surely, whether a child has a good idea or a good helper, *something* might be learned?

The question caught me like a fish on a long troll line, and I read through the theory, practice, and history of inventions. If I'd been an educator, I might have designed a course. As it was, I wrote two books, *Reinvent the Wheel* and *Build a Better Mousetrap,* that help kids reinvent classic inventions from Neolithic paint to a simple motor, using materials they can find around the house. (Ironic, isn't it, me and "hands-on" science books? But I knew if I could build it, anyone could.) Each invention project was preceded by the story of a problem the inventor faced and what inspired his or her solution, and was followed by an explanation of the science behind the device.[*] After dreaming up fifty reinventions, I exhausted my hands-on imagination, as well as the patience of my three resident testers, but not my curiosity about how we have

---

[*] For example, French physician René Laënnec was consulted in 1816 by a young woman who he suspected had heart disease. Putting his ear to her chest for a diagnosis was impossible, given her sex and age. He happened, however, to remember "the distinctness with which we hear the scratch of a pin at one end of a piece of wood on applying our ear to the other." He rolled a quire of paper into a cylinder, listened through it to her heart, and was delighted to find he could hear the beat more distinctly than if he'd put ear to skin. His stethoscope, as he named it, worked because the sound waves that would have dispersed in all directions were captured and redirected to his ear.

come to understand and manipulate the natural world. The historical moments when the explanation for a familiar phenomenon shifted from myth to fact, or when scientific truth was converted into a practical process or product, continued to fascinate me. I relish a good plot twist, and these discoveries or inventions often changed the trajectory of civilization. I went on to write books about the history and science of iron, gold, glass, ceramics, dyes, and greenhouse conservatories. But I encountered nothing more interesting than the series of discoveries that revealed the elegant workings of photosynthesis, the engine of nearly all terrestrial life. And, by the way, the invention of club soda.

fifteen

A Momentous Mint

According to ancient Greek myth, Apollo, pierced by one of Eros's golden, passion-inducing arrows, became enamored of the woodland nymph Daphne. Sworn to virginity, she rejected the god's advances, but he chased after her, pursuing her relentlessly across the countryside. As he gained on her and she neared exhaustion, she called out to her father, the river god Peneus, to save her from impending rape. Just as Apollo's hands reached to grab his prize, her skin turned into bark, her slender arms became branches, and her flowing hair morphed into leaves. She had metamorphosed into the beautiful bay laurel tree (and made Apollo the original tree-hugger). And that's about all the ancient Greeks had to say about foliage. It was an adornment, like hair. Only Aristotle imagined that leaves had any purpose at all, writing in *The Physics* that they exist to hide fruit from marauding birds and beasts.

Fifteen hundred years later, a medieval farmer, if asked what leaves are for, might have answered, in teleological fashion, that God created them to feed cattle, sheep, and goats: Otherwise what would they eat? Paracelsus, the Swiss physician and mystic who wrote and preached in the sixteenth century, believed God created leaves and other plant parts to heal man's illnesses, and left prescriptions in their shapes. Heart-shaped leaves were good for the heart; the liverwort's triangular leaves announced its ability to cure diseases of the liver.

In the 1640s, Belgian physician and alchemist Jan van Helmont ran an experiment he hoped would reveal the secret of what plants are made of. He potted a sapling willow tree in a tub filled with two hundred pounds of soil, and for five years added only water. At the end of the fifth year, he discovered the tree had gained 164 pounds while the soil lost an insignificant two ounces, and concluded that trees "arose out of water only." Van Helmont's experiment is one of the most famous in the history of science, not, obviously, for its conclusion, but because Van Helmont was the first to use quantitative methods to understand a living organism. What interests me, however, is that while Van Helmont was exacting enough to design a cover for the pot that kept "airborne dust from mixing with the earth" while also allowing evaporation from the soil, he didn't bother to weigh each year's fallen leaves. He didn't say why he omitted them from his calculations. I suspect he thought that leaves, doffed like clothes before a bath, were not an integral part of a tree.

Malpighi and Grew were well aware that leaves were generated from a plant's substance and assumed they had some function. In 1686, Malpighi cut off leaves from living plants, and found they grew less and produced fewer fruits. Leaves, he posited, serve "to allow the nutritive juice flowing from the fibers of the wood to be cooked . . . allowing growth of new parts." Their stomata allowed "excremental liquids" to escape. Grew didn't buy the poop hypothesis, but he couldn't decide whether the stomata were for evaporation "of superfluous sap, or the admission of air" for breathing. Fifty years later, Stephen Hales looked into the question, stripped plants of all their foliage, and discovered that denuded plants inevitably died. Like Malpighi, however, he concluded that leaves must be the plant's main "excretory ducts."

Enter, whistling, the polymathic genius Joseph Priestley.

Priestley was born in 1733 near Leeds, in the heart of northern England's wool manufacturing district. His father was a cloth-dresser, an artisan who singed, trimmed, and ironed rough woven cloth to transform it into finished fabric. When Joseph was six, his mother died giving birth to her sixth child. The boy had already spent much of his life at his grandfather's house in order to lessen the burden on his prolific parents, and now was virtually adopted by his father's childless aunt Sarah and her well-to-do husband. The adoption proved a mixed blessing. His aunt and uncle recognized his intelligence and saw to it that he was educated in a way his father could not have afforded. On the other hand, they were adherents of the strictest Calvinism. Joseph grew up believing that Adam's sin destined most of mankind, most likely himself included, for the roaring, stinking fires of hell. Only a few predestined "elect"—recognizable by their unwavering faith and sinless lives—would escape the eternal wrath of a merciless God. Joseph was terrified, especially because such horrors seemed no distant prospect. He was a sickly child with a disease, probably pulmonary tuberculosis, that often left him feverish and struggling for breath. He had watched his mother and a sister die, likely of the same disease, and so damnation must have felt not only real but imminent.

His illness came to a crisis when he was sixteen, as did his terror. He later wrote of this period that he was sure God had forsaken him, and experienced "such distress of mind as is not in my power to describe." He survived, however, and somehow in surviving, his belief in such an unforgiving deity vanished with his fever. He began to question not only the reality of original sin and other fundamental tenets of his family's religion, but even Christ's divinity. Decades later, this reevaluation would lead him to help found the English branch of the Unitarian Church and get him hounded out of England. At nineteen, although he was a long way from constructing his mature set of beliefs, he already found it impossible to either swear to his sect's ten articles of faith or affirm that he had experienced the requisite visitation from God and conversion experience.

His intransigence had important practical consequences. Like Grew, as a Nonconformist he was barred from attending Oxford or Cambridge. (This was in fact no great loss. By the mid-1700s, the universities' standards and enrollment had fallen sharply. For the most part, their diplomas were the equivalent of a passport issued to the sons of the Anglican gentry into careers as landholders, politicians, and curates. As one Cambridge historian writes, although it was possible for a prospective clergyman to get a decent, albeit circumscribed, education, it was equally possible to get a B.A. "and precious little else except advanced skills in drinking and driving a coach and pair.") Instead, thanks to the Act of Toleration of 1689, he could attend one of the excellent Nonconformist institutions, which at the university level were known as academies. Joseph, however, was in a bind. Because of his nonconforming Nonconformist views, he was unwelcome at most of the Calvinist academies his aunt and uncle had in mind for him. Luckily, Daventry Academy, an institution run by orthodox Calvinists who, most unusually, did not require a conversion experience and enrolled qualified men of a range of Protestant stripes, accepted him. It may have been the only institution that his aunt and uncle would pay for that their nephew could in good conscience attend.

The young man who arrived at Daventry was a tall-ish, thin, darting sort of fellow with an asymmetrical face, a kind heart, and a disconcerting stutter. He was academically advanced, at least in the course work he needed to become a minister, and was already proficient in Latin, Greek, Hebrew, French, and German and had started studying Arabic and the ancient Middle Eastern languages of Chaldee and Syriac. He had also learned enough logic, history, and philosophy—much on his own—to be excused from the first two of the five years of study. As for natural philosophy, he rejected the Calvinist idea that trying to unravel the secrets of the natural world was sinful pride, and considered it an honorable endeavor to appreciate God's creations. So, he plunged into courses on anatomy, mechanics, acoustics, and astronomy, struggled with mathematics (it would never be his strong suit), and independently read Boerhaave's *Elements of Chemistry* and Newton's *Opticks*. Fizzing with intellectual curiosity, he got up early, worked late, and learned shorthand so he could work with maximum efficiency.

Socially, the young man was in heaven. For a nineteen-year-old whose single childhood memory of fun was reading *Robinson Crusoe,* Daventry was a fairground of recreations. Emancipated from guilt and fear, he attended clubs and parties and even tried to woo the "cuddliest creature," a certain Miss Carrott. By the time he graduated, the anguished teenager who had once snatched a book of chivalric stories from his younger brother's hands and trembled with revulsion at the sound of an oath was transformed.

It was unlucky, therefore, that his first professional position or "calling" was as an assistant minister to an orthodox congregation of a hundred souls in the impoverished village of Needham Market in Suffolk, far from his native Yorkshire. He attempted a series of lectures on religious theory, but his parishioners quickly recognized his heterodoxy. With his Yorkshire dialect and his stutter, his preaching was an ordeal for all. His congregation diminished rapidly, and because Nonconformist clergy were paid by their congregants, so did his already meager salary. Then his brother (the one with the immoral book) told his aunt what a "furious freethinker" her favorite nephew had become, and she cut off his allowance. Priestley thought he might supplement his income by starting a school, but not a single pupil showed up. Only charity funds available to impoverished clergy allowed him to survive.

One resource he was rich in was friends, who were drawn to the openhearted young man. After three years in Suffolk, a sympathetic acquaintance rescued him, securing him a temporary appointment to a congregation in the town of Nantwich. Nantwich was in the tidewater salt-manufacturing region of Cheshire, both culturally and physically close to his hometown. His new congregants were more tolerant of his unorthodox religious views and could understand his accent. Without the pressure of a hostile audience and with perseverance, he improved his performance in the pulpit. Finally, he had escaped the gloom of orthodoxy, and his natural joy in the world seemed to bubble out of him. He couldn't keep himself from whistling in public or leaping over the counter at the greengrocer's, habits that amused rather than offended his new parishioners.

Best of all, when he started a small grammar school for the children of his congregants, he discovered a more congenial métier. A natural teacher, he was able to share his own delight in a subject, and there seemed to be no subject that did not delight him. At a time when grammar schools taught primarily Latin and Greek, he added math, history, and English. His scientific education was sketchy, but he spent long evenings talking with and learning from a Cambridge-educated vicar who lived nearby and kept up with the latest developments in the experimental sciences. With his small earnings, Priestley bought himself scientific equipment, including a microscope, a machine that generated static electricity, and an air pump, and shared the devices with his young pupils.

Priestley's success in the classroom led to a position as tutor at the Warrington Academy in Lincolnshire in 1761. Here he came into his own, taking on responsibilities at the school for hiring a chemistry teacher, preaching sermons at a local chapel (to highlight their unsanctioned nature, non-Anglican churches had to be called chapels or assemblies), and becoming a member of the board of governors of the town library. In June the following year, he married Mary Wilkinson, a well-read, assertive, chess-playing young woman who was the sister of one of his students. Mary, whom Priestley described in his memoirs as a woman "of excellent understanding, much improved by reading, of great fortitude and strength of mind, and of a temperament in the highest degree affectionate and generous," created a happy and busy social life for her still somewhat awkward husband. After a year of marriage their first child, Sally—named after his aunt, who was not appeased by this honor—was born.

Priestly had been pondering the educational needs of young men, particularly Nonconformists barred from government careers and comfortable Church of England livings, who needed to prepare for a world increasingly dominated by commerce and manufacture. He wrote a book on English usage, not for men writing speeches or sermons but for those who needed a guide to clear, everyday communication. Young men employed in "the useful arts," he believed, should be well grounded in modern history, mathematics, physics, and chemistry. While these insights were hardly his alone, he articulated them coherently in books, essays, and, notably, with detailed course syllabi.

Priestley also convinced a local surgeon to teach a course on "practical," that is, lab-based, chemistry at Warrington, and volunteered to act as his lab assistant. Yellow and blue flames, red fumes, acids that devoured metals, explosions, they all entranced him as they had generations of alchemists pursuing gold or the philosopher's stone. But he was also interested in the commercial side of chemical transformations, the processes by which clay turned into pottery, sand and lime into glass, and coal into coke.

In 1765, the University of Edinburgh granted Priestley an honorary degree for his work in education. That year, he embarked on a completely novel literary endeavor: writing a history of the experimental sciences. He decided to start with the new science of electricity, a subject that fascinated the public at the time. His position at Warrington had given him entrée into the intellectual community of nearby Liverpool, and one of his new friends helped introduce him to the "electricians," men like Benjamin Franklin, famous for his experiments with lightning and storing static electricity in Leyden jars; Englishmen John Canton, who verified Franklin's discoveries; and William Watson, who had passed an electric current along a wire across the Thames River. The scientists invited him to their periodic meetings in London coffeehouses and agreed to help him by explaining their work and lending him their papers. Even better, both for his book and science, they encouraged him to make his own explorations. Now thirty-three, Priestley threw himself into experimentation, melting wire with an electrical current and repeating Franklin's kite experiment (which kite he was wise enough to ground with a chain). He reported his progress in a flood of letters to his new friends, and barely a year later finished the seven-hundred-page *The History and Present State of Electricity, with Original Experiments*. The book won him election to the Royal Society.

He then turned his attention to another experimental science, "pneumatic chemistry" or the chemistry of air. Robert Boyle and his seventeenth-century contemporaries had considered air to be a single, homogeneous substance, one of the four "elements" of which the world was made. Any differences among particular samples of air—say, a human breath or the "mephitic" air that rises from certain springs or the bottom of swamps—was a difference in its condition. Just as meat could be good or "off," air could be good or bad. Of course, no matter what their current condition, meat was meat and air was air.

In 1757, a young Scottish physician named Joseph Black had sent a major tremor through the bedrock of that belief. While investigating a cure for bladder stones, he heated a sample of magnesium carbonate, and found it emitted an air that seemed distinct from ordinary air. This "fixed air"—"fixed" because it had been affixed to or stuck in the magnesium—is what we know as carbon dioxide, or CO_2. Black further realized that this air was the same one given off by burning wood, the exhalations of animals, and fermentation. Fixed air killed birds and small animals imprisoned in glass jars and snuffed out candle flames. When he burned crushed limestone (what we know as calcium carbonate), fixed air emerged and left behind quicklime (or calcium oxide). When he exposed the quicklime to common air, it turned back into limestone. Fixed air was both a distinct substance and a part of common air. This meant received wisdom was wrong: Air was not an element. Ten years later, English chemist Henry Cavendish produced another air, what he called "flammable air" (and we know as hydrogen).

What made these airs different from one another and how they related to common air, no one knew. But fixed air particularly interested Priestley, and he repeated Black's experiments for himself. After their daughter's birth, Mary convinced him to leave Warrington, where she feared the river air was undermining her own and Sally's health. Joseph found a position as a minister at Mill-Hill Chapel in Leeds, only seven miles from his birthplace. Because the minister's house was undergoing renovation, he temporarily took a house that happened to be next to a brewery. He knew the vats of fermenting grain produced an endless supply of this fixed air and, to his delight, the brewers let him experiment with it.

Priestley soon discovered that water could absorb the fixed air of fermentation, and that it then became bubbly and tasted slightly sour. By pouring water from one bowl to another in the flow coming off a vat (carbon dioxide is heavier than air, so it cascaded invisibly over the vat's edges), he "carbonated" it, creating an artificial Perrier or soda water. Priestley didn't try to capitalize on his invention, although he did propose it as a cure for scurvy and other illnesses. (It was Jacob Schweppe who patented bottled soda water in 1787 and, later, other carbonated drinks. Lest you think this unfair, although Priestley put the bubbles in the water, it was Schweppe who figured out how to keep them there. He put the soda water in an egg-shaped bottle that couldn't stand upright. As it leaned, the soda water contacted the cork, keeping it damp so it didn't shrink and allow the carbon dioxide to escape.) But if soda water was of little interest to Priestley, it did inspire him to consider an interesting scientific question. If fermentations, animals' respirations, swamp bubbles, and volcanic explosions all poisoned the air, how had God arranged for it to be "rendered fit for breathing again"? He understood there had to be some mechanism; otherwise, our air would have become hopelessly corrupted.

With these questions in mind and using laboratory equipment invented fifty years earlier by Stephen Hales, he continued to experiment with fixed air, enclosing insects, frogs, and mice in upside-down jars of common and fixed air until they lost consciousness. (A softhearted man, he often removed his subjects and was able to revive them before they died.) Then one day in the summer of 1771, he idly put a small mint plant into an inverted jar of common air to see what would happen to it. No one had tried the experiment, but then, why would they? If plants breathed, certainly they did so like other living creatures, by exhaling poisonous fixed air. Priestley assumed that his mint plant would die, just as mice did.

Not only did the mint survive in the jar; it continued to grow for months. Even more surprising was his discovery that, after all that time, "the air [inside] would neither extinguish a candle, nor was it at all inconvenient to a mouse, which I put into it." What was most astonishing is that if he put a plant in a jar in which a candle had guttered out or a mouse had suffocated, some days later a candle would burn and mice could breathe again.

Priestley repeated his experiments throughout the summer of 1771, to make sure he had made no mistake. He hadn't. Somehow, mint cured bad air. Thinking that perhaps mint was the only life-giving plant, he tried balm, foul-smelling groundsel (in case only sweet-smelling plants performed this restorative function), and spinach. They all worked. God had arranged it, he concluded, so that plants "reverse the effects of breathing." Priestley had the first glimmer that the living organisms and the nonliving environment are inextricably and reciprocally related—one of the essential principles of modern ecology. With remarkable intuition, he realized that the Earth's environment is a closed system. The paper he presented to the Royal Society on his findings on plant-generated, breathable air was a sensation, and he won the Society's Copley Medal in 1773. As Sir John Pringle concluded when presenting Priestley with the medal, "from these discoveries we are assured, that no vegetable grows in vain [and] every individual plant is serviceable to mankind; if not always distinguished by some private virtue, yet making a part of the whole which cleanses and purifies our atmosphere."

How plants accomplished their miracle of purification was uncertain. Did they take something noxious from the air or add something beneficial to it? Priestley had also discovered that plants placed in air "strongly tainted with putrefaction" often grew particularly vigorously. With this in mind, he picked the first option: The plant took something out of the air. What he thought it took out was "phlogiston," a weightless and invisible "principle" that was generally believed to depart a material when it burned or animals breathed it in. A mint plant, Priestley concluded, took phlogiston out of the air and left the air "dephlogisticated." What we know as oxygen was to Priestley "dephlogisticated air" or "good air," that is, air that had been purified of the phlogiston added by animals, decay, or fermentation.

From 1773 to 1777, Priestley focused on other scientific matters, as well as theological ones, but in early 1778, he again took up his research on plants and air. Reports had reached him that other scientists had been unable to reproduce his results, so he began repeating his earlier experiments. To his distress, this time his results were inconsistent. Sometimes his plants in jars produced good air and sometimes fixed air. Sometimes they barely altered the air at all. To clarify his muddled findings, he submerged little plants in vials of water, corked the vials, and analyzed the quality of the air that collected inside. He seemed at first to be making progress: The plants produced a particularly pure dephlogisticated air. But after he removed some plants from their vials, he noticed a "green matter" on the inside of the glass. To his astonishment, the green matter (which he assumed was neither animal nor vegetable) was releasing bubbles of good air.

He continued to experiment with vials of this interesting green matter, putting them near a stove and on sunny windowsills, and wrapping some in brown paper and afterward measuring the dephlogisticated air they produced. In the process, he arrived at his second great discovery: "As extraordinary as it will seem," he wrote, light is essential to producing good air. Unfortunately, this insight led him to wonder if "plants had not, as I imagined, contributed anything to the production of this pure air." Instead, he hypothesized, "light disposes water . . . to make a deposit of a greenish or brownish matter, and then to yield dephlogisticated air." It was all very confusing, and by 1779 he could write only that "upon the whole, I still think it *probable* [his emphasis] that the vegetation of healthy plants, growing in situations natural to them, has a salutary effect on the air in which they grow." Perplexed, he urged others to consider his results and investigate further.

sixteen

Leaves Eat Air

One of those who did was a Dutch physician, Jan Ingen-Housz. Ingen-Housz was born in 1730 in Breda, in the southern part of the Netherlands. His father was an educated man and the leading apothecary in town. Like Priestley, Jan had a gift for languages, and attended universities in Paris, the Netherlands, and Edinburgh. After completing his study of medicine, chemistry, and physics, he settled down to a career as a physician in his hometown.

His heart, however, was in experimental science. As a Catholic, he had no hope of a career at one of the Dutch universities, which were as off-limits to him as the Anglican universities were to Priestley. To his good fortune, an old connection opened a door for him. In the mid-1740s, the English army had helped the Dutch defeat a French invasion of Flanders, and the army had camped for some time

outside Breda. The chief medical officer of the English army
had been the young Dr. John Pringle. Pringle was fluent in
Dutch, and became a frequent visitor to the Ingen-Housz
residence, where he met the apothecary's precocious teenage
son. In 1764, Pringle—by then knighted, a physician to King
George III, and head of the Royal Society—invited the thirty-
four-year-old Jan to London and introduced him to London's
scientific and medical community.

Ingen-Housz found work at the London Foundling Hospital, where, among other duties, he inoculated the resident
children against smallpox, a terrifying disease that infected
more than a million Europeans annually, killing a third of its
victims. Many of those who survived were horribly scarred;
some were also blinded. Although by this time many physicians in England and the Netherlands recommended inoculation to their patients, there were few takers in those countries
and almost none in other European states. The practice was
banned in Paris. It is not hard to understand people's reluc-
tance, given the procedure, which involved a physician cutting a small vein and dropping in fluid or ground-up scabs
taken from a victim of a mild case of the disease. Besides,
while the procedure generally induced only a slight infection, it was not without risk: About 2 percent of those inoculated contracted full-blown disease.

Wealth and status provided little protection against
smallpox; the virus is highly infectious and can remain viable for months in the environment. By 1767, Maria Theresa,
empress of the Holy Roman Empire, and sovereign of a host
of other central European states, as well as parts of modern
Netherlands, France, and Italy, had lost two of her sixteen

children and several other family members to the disease. That year, at the age of fifty, she contracted the disease herself, and although she survived, she was deeply scarred. Her daughter-in-law, Empress Maria Josepha of Bavaria, died of the illness that same spring. Maria Theresa took one of her daughters to pray with her at the dead empress's unsealed tomb. When, days later, her daughter came down with smallpox and died, the empress blamed herself, although given the weeklong incubation period, she was certainly not responsible. The imperial court physician opposed inoculation (instead favoring bleeding and the ancient Japanese treatment of dressing the afflicted in red), but after two more imperial daughters were disfigured that fall, the empress overruled her doctor. She wrote to King George III to ask that a physician be sent to Vienna to inoculate her remaining family. It fell to Dr. Pringle to select the man.

The man who subjected the imperial family to a potentially lethal procedure would be taking a gamble. Success might make his career, but a failure could end it. Pringle thought of Ingen-Housz, both for his competence in the procedure and out of political considerations. If any of the empress's family were to fall ill, Catholic royalty would not have died at the hands of an English Protestant. Pringle approached Ingen-Housz, who accepted the job.

On his arrival in Vienna, the empress first had him inoculate twenty-nine commoners' children in her presence. When these human guinea pigs survived, he was allowed to treat the immediate imperial family, including the young Marie Antoinette. Again, everyone survived, and Ingen-Housz was honored, awarded a heap of gold ducats, and appointed

court physician with a lifetime pension. His chief responsibility over the coming years would be to inoculate dozens of royal wives, husbands, cousins, nieces, nephews, and
grandchildren at their villas and palaces across Europe—a
great improvement in working conditions over those at the
foundling hospital.

Ingen-Housz now had the opportunity to spread the
word about comprehensive smallpox inoculation, a cause he
fervently believed in. But welcome, too, was the time, status, and wherewithal he now had to pursue his scientific
interests. In May 1771, he joined Ben Franklin and two
American businessmen on a kind of applied-science tour of
the English Midlands, visiting marble-milling, silver-plating,
iron-smelting, and other manufacturing operations. He and
his companions also stopped in on Priestley, who demonstrated some of his electrical experiments, which so deeply
intrigued the Dutch physician he began his own electrical
investigations. When passing through Livorno, Italy, on the
Mediterranean coast, he hired a fishing boat and a crew of
eighteen, who captured five electric rays for him. In a makeshift lab on board, he attempted to correlate their size to the
intensity of the shocks they delivered, and to capture their
electrical charge in a Leyden jar. Then he dissected his subjects, with the hope (disappointed) of discovering where and
how they stored their energy. Naturally, he reported his findings to the Royal Society.

Ingen-Housz stayed current with Priestley's research on
air and plants, and in the summer of 1779 he took leave from
his post in Vienna, rented a villa near London, and set himself a discrete task. He would attempt to clarify the muddled

relationship among plants, sunlight, and air. Many of his experiments were similar to Priestley's, but his approach was fundamentally different. Priestley tested whole plants, but the Dutch physician used only leaves. Priestley had let his experiments run for days or weeks; Ingen-Housz completed each test in a matter of days.

After completing 546 experiments in ninety days, Ingen-Housz had answers. Priestley had been right the first time. Leaves and other green parts of a plant turn fixed air into good, dephlogisticated air. Plants require sunlight to carry out the transformation. Moreover, Ingen-Housz proved that the brighter the light is, the more dephlogisticated air is produced. The good air emerges primarily from the stomata on the undersides of leaves. That "green matter" in the vials, he demonstrated, was vegetal in nature and also produced good air. Roots, flowers, and ripe fruit always produce only fixed air. Most remarkably, in the dark *all* parts of a plant, including leaves, produce fixed air. Although the amount emitted in darkness, he calculated, is only a hundredth of the amount of good air emitted in two hours of sunlight, he wondered if plants should be removed from a sickroom at night.

Ingen-Housz was able to explain Priestley's inconsistent findings. The English minister hadn't understood to what degree variations in the amount of sunshine and the changing daylight length affected the output of good air. Nor did he realize that the gradual senescence of leaves would reduce the output of good air. In addition, he hadn't taken into account how root systems altered his experiments: the greater the mass of roots, the greater the production of fixed air. To put it in modern chemical terms, the amounts of carbon dioxide and oxygen in Priestley's closed containers had fluctuated, depending on light intensity and duration in his lab, the time of day that he tested the air, the ratio of roots to leaves of a given plant, and the ever-changing size and health of his experimental subjects.

Ingen-Housz also had a far greater appreciation of painstaking lab procedures. Experiments, he realized, had to be repeated in precisely the same fashion and under exactly the same conditions before any valid conclusions could be drawn. Sloppy lab procedures would have measurably influenced Priestley's results and those of others trying to reproduce his experiments.

Ingen-Housz discovered which parts of a plant produced good air and fixed air, but it was Jean Senebier, a Swiss pastor, librarian, and amateur botanist, who proved that the two gases were related. In 1781, he repeated his predecessors' experiments, ran his own, and wrote up his results. The pastor's prose was in want of a hedge trimmer (he wrote 2,100 pages on his plant experiments), but he made a major advance: in order for leaves to produce good air, they must have a supply of fixed air. To account for the difference between the two airs, he posited that phlogiston was liberated from air and incorporated in plants' new growth.

Even while Senebier was imagining a new role for phlogiston, the concept was under attack by Antoine Lavoisier. Lavoisier was a brilliant French chemist and the youngest person ever elected to the French Academy of Sciences. Burning, according to Lavoisier, has nothing to do with a supposed phlogiston, but with an element—a term he redefined as a chemical substance that could not be further broken down into other chemical substances—that he named *oxy-gène*. Oxygen readily combined with many other elements; this process of combination, or *oxidation,* could occur rapidly with a sudden release of heat and light, as when wood is heated, or slowly, as when iron rusts.* Proof of his theory lay in the fact that while rusted iron appeared to involve a loss of a substance, it actually weighed more than untarnished iron. In fact, the mass of rusted iron was exactly equal to the iron plus the mass of incorporated oxygen. Modern chemists had to be bookkeepers, he wrote, measuring precise volumes and weights of liquids, solids, and gases, keeping track of inputs and outputs that always balanced.

Lavoisier presented his revolutionary chemistry to the Academy in 1778 and published his masterwork, *Elements of Chemistry,* in 1789. Priestley never accepted Lavoisier's "new chemistry," and Ingen-Housz accepted only a part. But by 1796, Senebier was a believer, adopting the new concepts and terminology. Leaves, he now asserted, absorb that small portion of air that is carbon dioxide and, in sunlight, decompose it into carbon and oxygen. The carbon from carbon dioxide becomes the organic matter of plants, and oxygen ends up in the atmosphere. He didn't have the full story yet, but he was the first to put photosynthesis in modern chemical terms.

Nicolas-Théodore de Saussure, born in 1767 in Switzerland, was a young man when Lavoisier's book appeared. His

---

* Lavoisier used the term *oxidation* to describe reactions in which an element combines with oxygen. Later, oxidation was redefined as any reaction in which an atom loses electrons in its outermost shell to another atom, thereby becoming positively charged.

father, Horace-Benedict de Saussure, was a professor of natural sciences at the Academy of Geneva, the man who coined the New Latin word *geologia,* discovered fifteen minerals, and became a famously intrepid mountaineer. He tutored his eldest son at home and brought the boy along on his scientific treks through the Alps. These adventures were so rugged and at such heights—the purpose was to study the relationship of altitude to the density of air, among other puzzles—that one time their porters threw away the expedition's provisions in a desperate effort to convince the pair to descend. The unusual education suited Nicolas-Théodore, who found his calling in the new chemistry, although it perhaps had a less beneficial effect on his social development. He grew up to be such a painfully shy, self-effacing man that while his work won him an early professorship at the Academy of Geneva, he was never able to stand before students and lecture. He was a masterful experimenter, however; a double-entry chemical bookkeeper par excellence. Thanks to the better instruments of his day and his extraordinary skill in using them, he was able to measure the volume and weight of gases exchanged by plants to the hundredths of an ounce.

Saussure's passion was the chemistry of plant physiology, and he put together the modern description of the basic process of photosynthesis. He demonstrated that, in sunlight, fixed carbon appears in leaves simultaneously with the disappearance of carbon dioxide and its replacement by oxygen in the surrounding air. Moreover, the green parts of plants actively transform the carbon into organic matter. In other words, leaves *eat* air. When plants grow larger, he discovered, atmospheric carbon dioxide is by far the largest source of that increase. Minerals from the soil—which were still assumed to provide the bulk of new plant material—actually comprise well less than 5 percent of a plant's mass. It was Saussure who explained why in darkness all parts of a plant release carbon dioxide. Plants—like animals—respire, meaning they use oxygen to burn carbon-based sugars to fuel growth, concoct scents, manufacture insect-killing resins, and carry out other essential activities. In essence, roots and seeds "breathe" as we do. In fact, even leaves respire a little bit during the day. It's just that they take up so much more carbon dioxide in sunlight than they release through respiration that their respiration is easy to overlook.

The Swiss chemical bean counter was able to make such minute and precise measurements that he discovered an error in the chemical books. The mass of the carbon taken in by leaves and the mass of minerals taken up by roots did not equal a plant's total increase in mass. A plant weighs more than the sum of the carbon and minerals. The extra weight, he realized, comes from water.

Van Helmont had been right, or at least a little bit right, after all: Plants *are* made of water. Although 99 percent of the water rising from a plant's roots exits its leaves as water vapor, about 1 percent is incorporated into the substance of a plant. It is a minuscule but crucial amount. H_2O, split apart by the energy of sunlight, contributes the hydrogen for those molecules of $C_6H_{12}O_6$ (the simple sugar, glucose) that plants make and use for energy. And although Saussure didn't know it, the oxygen produced by plants, the oxygen that all multicellular life on Earth requires, comes from the oxygen in water, not, as everyone assumed, the oxygen in carbon dioxide. Remarkably, this last fundamental fact of photosynthesis was not uncovered until the 1930s by Stanford professor C. B. van Niel.

Saussure published his masterwork, *Chemical Investigations of Plant Growth,* in 1804, the year that Joseph Priestley died. In a thirty-year span, the role of leaves had been transformed from incidental to essential. The remarkable process of photosynthesis—the creation of organic compounds using light energy—was revealed.

Understanding exactly *how* sunlight transforms carbon dioxide and water into carbohydrates was another matter.

seventeen

The Vegetable Slug

The inch-long sea slugs that I have come to see in Dr. Sidney Pierce's lab at the University of South Florida are, unlike the usual brown slugs, lovely animals. Hovering in the clear water of a ten-gallon aquarium, they look like translucent scraps of bright green, ruffle-edged arugula. One lettuce bit ripples its edges on the front glass, and I can make out under the white light of a fluorescent bulb what looks like a pair of knobbed green horns. This slug is adorable. It's slug à la Pixar.

The aquarium is empty except for Pierce's little herd. Or should I say little crop? For these creatures, perfectly healthy and presumably happy, live like plants by photosynthesizing. For the past eight months, they have eaten nothing but the photons pouring from the lightbulb above. Remember the borametz, the mythical vegetable lamb chimera said to live on the Caucasian steppes? This slug, *Elysia chlorotica,* and a few related species are its real-life cousins. They are genuine chimeras, part sea slug and part algae.

I am entranced by the dance of the lettuce leaves, but I know Dr. Pierce has limited time. We adjourn to his office next to the lab. He tells me he's thinking about retirement, but with a boyish mop of white hair, a rosy complexion, a restless physical energy, and a certain provocative attitude—taped to his office door is a warning to any students coming to inquire how they might raise their grade that the only way to do so is to actually do better on his exams—it's hard to credit. A "biochemist by trade," he is also a retired chairman of the biology department at the University of Maryland, an active professor at USF, and author of a multitude of articles on the biochemistry of invertebrates, as well as a textbook on invertebrate anatomy. He has also become a "sea monster" expert. From time to time, a huge gelatinous blob washes up on a beach somewhere in the world. When the discoverer inevitably wonders if this is evidence of some previously unknown creature of the deep, he or she is often directed to Dr. Pierce. Just as inevitably, Pierce pronounces the subject to be something less than marvelous, usually a piece of decomposing whale blubber. There is a certain irony, I think, in his fame as marine mythbuster, given that he has spent the past two decades elaborating the physiology of what would seem to be a scientifically impossible creature, a photosynthesizing animal.

Pierce's office is in a state of confusion. He is in the middle of packing up his papers so he can adjourn to Woods Hole Marine Biological Laboratory on Cape Cod, Massachusetts, for his annual summer research. It was at Woods

Hole, he tells me, about twenty-five years ago that a couple of graduate students brought him the first specimens of the photosynthetic *Elysia chlorotica*.

"Until these students brought them in, no one knew they lived off Cape Cod, in spite of the fact that this big fancy lab had been studying the local marine environment for something like a hundred years. I was studying salinity tolerance at the time. Oddly, these completely marine creatures turned out to be some of the most salt-tolerant creatures I'd ever encountered, able to survive in everything from fresh water to water as salty as the Dead Sea. This was pretty remarkable since a sea slug is basically an unprotected ball of slime, crawling around in a marsh. So, I featured them in the talks and slide shows I gave about our research on salinity tolerance. For years, after my talk, people would come up to me and say, 'Very nice talk, but look at those slugs! Those slugs are green. Why aren't you studying that, you fool?' Finally, I listened."

Pierce wasn't the first to work on green sea slugs. In the 1960s and '70s, a group of scientists studied a European species, wondering if its coloration meant it used the sun's energy to photosynthesize. (In general, green animals—fish, reptiles, amphibians, and birds—look green because blue light reflects through a layer of yellow pigment.) If the European sea slug did photosynthesize, its color must derive from green chloroplasts, which are the *organelles* (organs of a cell) in plants that use solar energy to transform carbon dioxide and water into glucose. The scientists found that the slugs did have chloroplasts, an unprecedented innovation in an animal. How these little engines of photosynthesis came to be inside an animal was an interesting mystery, and they started out to investigate.

Their first step was to remove the chloroplasts from the slugs.

"Of course, they immediately ran into the mucus problem. Sea slugs," Pierce explained, "are enormously capable of making mucus. They're practically nothing *but* damn mucus. Put a sea slug in a blender to grind it up, and when the little packets of anhydrous mucus that lie right beneath its skin hit water, they explode into this big wad of snot. Out of which you can centrifuge nothing.

"Before they could solve the mucus problem, the team discovered *Symbiodinium,* which are algae that live symbiotically—the whole alga that is, not just its chloroplasts—inside some invertebrates, like coral and giant clams. Coral reef bleaching was becoming a big issue, and these guys figured out that when coral polyps become stressed, their symbiotic algae flee. The polyps then no longer have access to the carbohydrates and amino acids that the algae produce, and they bleach and die. The researchers got enormously famous for this work, and gave up on the sea slugs forever. The green sea slug business sat fallow for years until I backed into it."

The first thing Pierce and his graduate students had to do was to tame the mucus. Running through a list of compounds developed to treat cystic fibrosis, a human disease that causes thick, sticky mucus to build up in a sufferer's lungs and digestive tract, they found one that also keeps the molecules of slug mucus from clumping together. They were then able to homogenize the sea slugs, centrifuge the goop, and isolate their chloroplasts.

The bulk of a chloroplast is filled with *thylakoids,* which look like stacks of flattened, green sacs, and *stroma,* a gel-like substance that surrounds them.

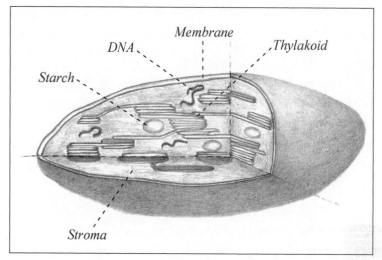

A chloroplast.

The outer surface of the thylakoids is where the first phase, the *light-dependent* phase, of photosynthesis begins. Attached to the thylakoid membrane are pigments—most prominently, chlorophyll—that act like antennae tuned to receive the energy of visible light. (Chlorophyll absorbs red and blue wave frequencies and reflects those in the green range, which is why leaves look green to us.) During daylight hours, when photons hit these pigment antennae, they bang off a flow of electrons. Some of those electrons wind up in the chemical bonds of ATP, that temporary energy storage molecule. Others split water into its hydrogen and oxygen components. Some of that hydrogen is incorporated in NADPH, another energy storage compound. As for the oxygen, some is used by plant cells, just as it is in animal cells, to fuel the breakdown of stored sugars. Most slips off into the atmosphere, where, to our great good fortune, we can inhale it.

Plant cells use ATP and NADPH for their immediate needs, like moving chemicals across cell membranes. But some of these molecules are used in the second half of photosynthesis, the *light-independent* phase, which takes place in the stroma. In this phase, which can occur both during the day and night, the energy molecules are used to combine hydrogen with carbon dioxide to make glucose, or $C_6H_{12}O_6$. (You may recall this from high school biology as the Calvin cycle. I didn't.)

Some of the glucose is converted into another sugar, the less chemically reactive sucrose, or $C_{12}H_{22}O_{11}$. Sucrose travels through the phloem headed for use in creating new cell components or for storage in, say, sugarcane stalks or fruit. String a hundred or so glucose molecules together and you have starch, which is stored in parenchyma throughout the plant— but most dramatically in tubers like potatoes and carrots— and is easily oxidized when needed for plant growth. In some plants, sucrose is converted into fatty acids and stored in oil-rich seeds, including those of the sunflower, rapeseed, and peanut. Weave several hundred to thousands of glucose molecules together, and you have cellulose, the tough matter of cell walls and what the cereal box notes as "dietary fiber."

Animals can't photosynthesize. We are missing two major components of the process. One, we have no chloroplasts. Two, we lack the genes that instruct the creation of the enzymes—most notably RuBisCO—that catalyze the chemistry of photosynthesis. Consider photosynthesis to be a pinball machine. Even if we animals somehow acquired the cabinet (the chloroplast with its stroma and thylakoids), we would still lack the flippers, bumpers, and springs (the enzymes) as well as any written instructions (the DNA) to play a game.

Which brings us to the question: How do the chloroplasts in the cells of green sea slugs manage to function? First, Pierce tells me, his sea slugs must eat at least one meal of algae before they can photosynthesize. The algal species they prefer, *Vaucheria litorea,* looks like translucent green straws. A newly hatched, brown slug crawls up to a straw and, sipping on it like a child drinking from a juice box, siphons the *Vaucheria*'s chloroplasts into its mouth. The chloroplasts then take up residence in the slug's gut cells, which form a network of ducts throughout its body. One meal of *Vaucheria* and the slug turns green and never needs to eat anything—except sunshine—again.

If slugs lived for only a few days or even a few weeks, Pierce says, those ingested chloroplasts might survive long enough to capture all the solar energy they need. But the slugs live nine months or longer, and the light-harvesting apparatus of the thylakoids, constantly bombarded by photons and stripped of electrons, wear out and need regular repair. Chloroplasts in plants and algae like *Vaucheria* have some of their own DNA, DNA that is different than the DNA of the plant or alga itself. The chloroplast's DNA directs the production of some of the chloroplast proteins. But a larger portion of the DNA that instructs chloroplast operation, including the repair instructions, resides in the DNA of the plant or alga. This is why chloroplasts cannot live outside a cell. They rely on their own DNA *and* the DNA of the organism in which they live. So, how do the thylakoids in sea slug chloroplasts manage to get repaired?

One possibility Pierce had to consider was that no repair was needed. Maybe sea slugs don't repair worn-out chloroplasts, but instead hold some chloroplasts in reserve to fill in for losses. To test this hypothesis, he set them swimming in a bath of radioactive amino acids, which are the building blocks of proteins and enzymes. After several weeks, he found the sea slugs were "hot." The slugs had manufactured new spares from the radioactive amino acids using instructions encoded in its own DNA. *Somehow, algal DNA has been incorporated into the slug genome.* In fact, Pierce and his students ultimately proved that the slugs can manufacture the entire sixteen-enzyme pathway to make chlorophyll. He also demonstrated that unhatched sea slugs contain the genes for photosynthesis, which means the slugs' ability to make chloroplast proteins is inherited. All they need is a first meal of chloroplasts to jumpstart the process.

No one had ever demonstrated a transfer of functional, heritable genes between taxonomic kingdoms—Plantae and Animalia, in this case—before. Pierce and his team have painstakingly identified the specific algal genes permanently embedded in the slug genome, a process that Pierce says has been "like trying to find a needle in a haystack without knowing what the needle looks like." Pierce's work has been reproduced by other researchers and there is little question that, however improbable it may seem, his sea slugs are aquatic chimeras between plant and animal. It's as if someone proved the existence of the borametz.

It occurs to me that photosynthesizing might be a useful trait for humans as well as sea slugs. I have very pale skin, a liability in our era of ozone depletion, and I sunburn quickly.

What if I could incorporate algal DNA into my genome and manufacture chlorophyll in my epidermis? I could bask in the sun without burning. Better yet, I wouldn't have to bother with cooking dinner. Or, I could store those sugars, burn them on cold winter nights, and save on the gas bill. I would give "going green" new meaning. I pose the question to Pierce.

"Dream on." He laughs.

Later, I look further into the question. Enough solar energy falls on the Earth in one hour that, if just we had the technology to capture it, we could fuel all human energy needs for a year. Plants turn only a tiny portion of that energy into biomass, in part because chlorophyll and other plant pigments trap light only within a narrow bandwidth of 400 to 700 nanometers. (More than half of solar energy arrives in the infrared and ultraviolet range and is useless to plants.) And although photosynthesis is a spectacular feat and powers almost all life on the planet, it is inefficient. Leaves convert only about 5 percent of the solar energy in its bandwidth into stored energy. Photosynthesis therefore means using a large, horizontal light-capturing surface area—that is, lots of leaves—to power a small volume of living cells. What about trees? you may wonder. Don't be misled by their size. Only about 1 percent of their vast bulk is living, energy-consuming cells.

I compared myself to a magnificent old oak down the street. Estimating that the tree has about a fifty-foot-wide canopy and using a formula suggested by Dr. Kim Coder of the University of Georgia, I estimate that the tree has about eight thousand square feet of leaves—nearly a fifth of an acre—to power its living cells. If I were to arrange for an algal transfer into my epidermis, I would have less than twenty square feet of photosynthetic skin, of which only half could I expose to the sun at any one time. On the other hand, I am composed of a relatively substantial volume of mostly living and hardworking cells, both my own and billions of resident microbes. If a green me sunbathed continuously between 9:00 A.M. and 3:00 P.M., I would collect about ninety kilocalories. Even given my sedentary existence, I figure I burn some 1,400 kilocalories a day. The bottom line is that for me to survive by simply soaking up rays I'd need a day with ninety hours of sunlight.

So, there's a good reason that plants have no brains, gather their water and food without moving a muscle, and don't go in for Kama Sutra. If any animal could live by photosynthesis, it is not surprising that an inch-long, translucent slug—whose common name derives from *sluggard*—is the one to manage the trick.

eighteen

Once in a Blue-Green Moon

It is no accident that the chloroplasts in Pierce's slugs—and chloroplasts in algae and plants—have their own DNA. The ancestors of chloroplasts are cyanobacteria (also known, confusingly, as blue-green algae), an ancient group of bacteria that live independently, floating around on the surface of the world's oceans, subsisting on readily available sunlight and carbon dioxide and reproducing by fission.

About three billion years ago—or 1.5 billion years after Earth formed and 300 million years before the cyanobacteria appeared—the planet looked radically different than it does today. The new continents, far smaller than the ones we know, were mostly submerged, with just a few barren outcroppings. The ocean was a rich green color, thanks to massive amounts of dissolved iron, and it was the temperature of a hot bath. The sky was hazy and orange with carbon dioxide, ammonia, and methane that spewed from active volcanoes. There was no free oxygen on the planet, in either the water or the atmosphere. All atoms of oxygen were bound up with other elements, mainly with hydrogen in water and carbon in carbon dioxide. Deep in this strange ocean were single-celled bacteria and archaea that lived in the hot, mineral- and gas-rich vents on the ocean floor. They made the energy to survive by reacting elements together in a membrane and using some of the energy released for their own simple metabolic needs. Some of these single-celled beings (*prokaryotes,* pronounced pro-CAR-ee-oats) were bacteria that stole electrons released in the combination of sulfur and iron abundant in those primal waters. Others, especially the archaea, corralled a little energy in reacting the hydrogen escaping from the Earth's core with carbon dioxide to make methane.

About 2.7 billion years ago, a new kind of bacteria evolved. These floated near the surface of the green ocean, and instead of collecting energy by reacting various chemicals together, they used the energy of sunlight—the photons streaming toward Earth—to strip off electrons from various chemical compounds in the waters around them. Some of these photosynthetic bacteria stripped electrons from hydrogen sulfide (H_2S), others favored hydrogen molecules (H_2), but the ones we care about—all the species of cyanobacteria—snagged electrons by splitting water (H_2O).

Cyanobacteria converted that electron energy into ATP. They then used up the ATP in fastening hydrogen protons onto readily available molecules of atmospheric carbon dioxide, thereby making sugars. The sugars became the building blocks of the rigid wall that separated a cyanobacterium from its watery environment, as well as its thick, mucosal extracellular coating. That coating was critical for cyanobacteria living at the ocean's surface. Ultraviolet light was as yet unfiltered by an ozone layer and would have otherwise fried their DNA. Whenever a cyanobacterium split H_2O, it burped a tiny bubble, an infinitesimal O of oxygen into the water.

Cyanobacteria lived an easy life. They had no need to chase down prey: They floated at the boundary of their two "foods," water and atmospheric carbon dioxide, and their energy supply was boundless. They divided and redivided and populations doubled and redoubled, all the while splitting water, fixing carbon dioxide into cell membranes, and popping out pinpricks of oxygen into the water. The numbers of cyanobacteria became incomprehensibly large, but know this: A single bacterium that starts dividing when the sun rises at six o'clock in the morning can, under perfect conditions, become a population of more than 34 billion by the time the sun sets. As the hundreds of millions of years passed, cyanobacteria became so abundant that they formed slimy, floating sheets in the open oceans. In shallow waters off the barren continents, the mats piled up and, interleaved with thin layers of mud and dead bacteria, formed pillows and domes and great reeflike structures called *stromatolites* that in shallow waters rose above the surface. If you had been able to scan the horizon 2.5 billion years ago, anything you saw that wasn't water or barren rock was stromatolites. In deeper waters, the mats accumulated in cones and columns as much as a hundred feet high.

The oxygen that cyanobacteria produced immediately

united with the vast quantities of iron dissolved in the oceans. Slowly, the oceans rusted, sending billions of tons of iron oxides to settle on the ocean floor in layers as thick as a half mile. (All the iron ore we mine today was formed in this era.) Finally, about 2.2 billion years ago, all the free iron, as well as other oxygen-hungry metals in the oceans like sulfur and manganese, had been oxidized. For the first time, oxygen bubbled into the oceans and floated off into the atmosphere, and the "Great Oxidation Event" began. The atmosphere gradually cleared, the sky and water became the blue we see today, and a protective ozone layer developed in the stratosphere. Many of those prokaryotes that had made their living by reacting iron and sulfur went extinct; other species were poisoned by free oxygen and either died out or were driven to the anoxic ocean depths. But those bacteria that reacted oxygen with carbon-based molecules thrived.

It was not, however, the single-celled bacteria that made the most of the new fuel. Some time between three and two billion years ago, another type of microscopic creature, a *eukaryote* (yoo-KAR-ee-oat), appeared in the oceans, likely descended from a one-time fusion between two single-celled individuals. Unlike a single-celled bacterium or archaeon, a eukaryote has a nucleus, which has its own membrane that encloses paired, threadlike chromosomes. Eukaryotes reproduce in a more complex way, by creating (through *mitosis*) two identical daughter cells that, by halving their genetic material (through *meiosis)*, produce gametes with only one chromosome. The gametes from two different individuals then fuse to form a new individual with a mix of its parents' genes. This sexual reproduction turned out be a boon for diversity, and that ur-eukaryote generated all multicellular creatures past and present, including, ultimately, us.*

Eukaryotes had another new feature. Unlike bacteria that have a rigid cell wall, eukaryotes' cell walls are made of a flexible, dynamic network of fibers. Flexible walls meant eukaryotes could stretch and bend to envelop other creatures and devour them. One day, about two billion years ago, a eukaryote engulfed a particular bacterium (probably related to the one that causes typhus) that made its living by using oxygen to metabolize sugars. The eukaryote should have deconstructed the bacterium. But for some reason, and just this once, this bit of dinner proved indigestible. Not only did the prey survive inside the predator; it reproduced, and its descendants survived inside the host's descendants. A symbiosis developed, with the bacterium reacting with oxygen and sugars provided by its host and the host snaring the resulting energy. The relationship, albeit born of an unsatisfactory dinner date, proved enduring and enormously productive. The engulfed bacteria became mitochondria, organelles that are the internal combustion engines of multicellular life.

Next, about 1.6 billion years ago, one of these oxygen-burning eukaryotes bumped into and engulfed a cyano-

---

* Much is unknown about the origins of eukaryotes, and there are many competing theories. The fusion could have been between two bacteria or, more likely, between a bacterium and an archaeon. Recent gene sequencing raises the possibility that eukaryotes may also have genes from an organism in a third single-celled kingdom that has left no modern descendants. This highly unlikely merger of three organisms could explain why the eukaryote evolved only once and all multicellular life is descended from it. Today, all multicellular eukaryotes are capable of reproducing sexually via mitosis and meiosis. (Dividing cells by mitosis is also the way a eukaryotic organism grows.) When, how, and why sexual reproduction developed is debated, and the story of eukaryote evolution is far from certain.

bacterium. This time, and just this one time, it was the cyanobacterium that survived the eukaryote's digestion. No one knows exactly how this happened, but there is recent evidence that a third party was involved. The research of Rutgers University biologist Debashish Bhattacharya indicates that another individual showed up for this particular dinner party, a *Chlamydia*-like bacterial parasite, and was simultaneously engulfed. This unique convergence allowed the cyanobacterium to survive inside the predator. The parasite did not survive, but certain of its genes became incorporated in the genome of its host. (This is not as unlikely as it sounds. Bacteria readily exchange genes with one another. This "lateral gene transfer" is part of the reason bacteria can evolve so quickly to thwart our antibiotics.) The parasite's genes created a crucial conveyor belt that whisked the sugars produced by the cyanobacteria to the host cell. The cyanobacterium not only was spared, it was able to reproduce. Its descendants survived inside the eukaryote's descendants, catching sunlight, building carbohydrates, handing them over to the host, and burping pinpricks of oxygen. Over time, some of the cyanobacteria's genes transferred into the host's nucleus, and eventually so many transferred that the guests could no longer survive independently. The cyanobacteria had become permanent residents, the chloroplasts.[*]

---

[*] Shed no tears for that former free-range cyanobacterium. Its descendants are found in all members of the plant kingdom, which now covers 75 percent of the earth's land surface, from deserts to tundra, and constitutes 90 percent of the world's biomass. As for the descendants of free-living cyanobacteria that didn't shelter inside eukaryotes, they have diversified into more than six thousand species and occupy nearly every watery or damp niche on the planet. Altogether, cyanobacteria contribute half of the new biomass—55 billion tons—produced on Earth every year.

Some of the early seafaring photosynthetic eukaryotes, or algae as we can now call them, found a particularly salubrious niche in the freshwater shallows of river deltas and bays. Here, they found a good supply of mineral nutrients that weathered out of rocks on the nearby shore. Fast-forward to 500 million years ago. (When we talk in terms of millions instead of billions of years ago, doesn't it feel almost like yesterday?) Some freshwater algae dropped a few fragile filaments down to the wet sand and prospered by anchoring in a harbor of plenty, instead of drifting in and out with the wind and tides. The algae reproduced and developed into colonies that floated on the water's surface in sheets. They also formed a symbiotic partnership with fungi (see chapter 10) that had colonized land earlier. The anchored algae proliferated and diversified. About 450 million years ago, some of them moved fully onto land, becoming the progenitors of the two major divisions in the kingdom of Plantae.

One division, the bryophytes, includes today's liverworts, hornworts, and mosses. Bryophytes maintain close contact with the ground—nearly as close as algae does to water—because, like algae, they have no means of moving water up and around their corpus. They also need a damp environment because their sperm must swim to fuse with egg cells. The resulting organism grows to form a structure—the sporophyte—that then releases airborne spores. The other division, which encompasses all other modern land plants, is the tracheophytes. As the name suggests, tracheophytes have tubes that conduct fluids throughout the plant.

The earliest tracheophytes also kept a low profile.

Although they could move water internally, their cells had no external coating to protect them from desiccation. (None had been needed in the watery environment from which they'd emigrated.) To grow tall, therefore, would be to risk death from drying winds. But if lying low was an excellent strategy for survival, it was not so good for reproduction. Like the bryophytes, the early tracheophytes all reproduced with spores, and the greater the height from which an individual could launch its spores, the greater likelihood of its spreading its genes the most widely. So, the little vascular plants began to evolve a clear, waxy cuticle. As it turned out, not only were taller individuals more likely to have more offspring, they were also more likely to reach maturity to reproduce: No neighbors put them in the shade. The race of the tracheophytes toward the sun was on.

A clear cuticle released the potential of tracheophytes. It kept internal water from evaporating while simultaneously allowing light to reach the chloroplasts. A completely impermeable coating, however, would have prevented the plant from absorbing atmospheric carbon dioxide. The solution was pores. And, wouldn't it be nice if those pores could open and close? Opened, the pores would maximize carbon dioxide intake. Closed—say, in the desiccating heat of midday or at night, when the plant couldn't photosynthesize—they would minimize water loss.

In fact, one of the earliest tracheophytes, the long-extinct *Cooksonia,* was a delicate, inch-high species equipped with just these features. *Cooksonia* was composed of a single, slender green stalk that bifurcated into two green upright and leafless stems. The stalk and stems were pocked with stomata.

Each stomate was surrounded by two sausage-shaped "guard cells" that, expanding and contracting by filling and purging water, altered the size of the pore. This tiny plant emerged about 425 million years ago and proceeded to blanket great swaths of Euramerica, the ancient amalgamation of what is now Europe and North America. By 400 million years ago, *Cooksonia* had been joined by similar species, some as tall as two feet and more elaborately branched. The plants tended to grow in dense patches where, packed tightly together, their fragile stems supported each other. Tiny mites, centipedes, and the quarter-inch precursors of modern spiders scooted along in this miniature landscape.

The long-extinct *Cooksonia*.

None of these plants had leaves. It wasn't that *Cookso-nia* and its companions couldn't make leaves. Leaf-making didn't require a once-in-history fusion or a fantastically rare undigested bacterium. The principal genes responsible for making leaves are the same ones involved in making branches, meaning that the ability to make leaves is as old as plants themselves. No, these plants were leafless because they had no reason to spend precious carbohydrates making and maintaining leaves. Earth's atmosphere was brimming with carbon dioxide, at four thousand parts per million, ten times more plentiful than today. A few photosynthesizing stems were sufficient for satisfying a plant's need for carbon. Leaves would have been a liability, capturing too much solar energy in this hothouse climate and causing the plant to overheat. Even if hypothetical leaves had been riddled with stomata, the roots of this era were still mere filaments, too attenuated to transport the volume of water needed for evaporative cooling from lots of leaves. So, *Cooksonia* and its cohort went unclothed.

The sticklike plants of the era were unimpressive to look at, but they were gradually making an impact on the ground beneath them and the air above them. Their roots may have been thready, but in harness with mycorrhizae, they continually scratched away at rock, liberating minerals molecule by molecule. Communities of aerobic microorganisms flourished, feeding on the dead roots and fallen stems of the new ground cover. The carbon dioxide these microbes exhaled, trapped in a new layer of humus and mixed with rainwater, became carbonic acid. This mild acid further accelerated the weathering of rock. Simultaneously, calcium and magnesium silicates, newly exposed to air, reacted with carbon dioxide and slowly, slowly pulled billions of tons of the gas from the atmosphere, sequestering it in organic compounds. From the start of the Devonian era about 415 million years ago to its end 360 million years ago, the planet's CO_2 level plummeted by 60 percent to 1,600 ppm. As it fell, the greenhouse effect weakened, and the planet cooled.

As the atmospheric carbon dioxide level fell, plants could safely capture more solar energy without overheating. The tiny, forked plants that had spread far and wide developed more branches, thanks to mutations in a class of genes known as KNOX. When KNOX is expressed, a stem grows straight. When the gene is silenced, the stem produces a sideways growth. As the environment changed, mutations that turned off KNOX periodically were favored by natural selection. Plants with additional photosynthetic branches trapped more energy, seized more carbon dioxide, and converted it to more plant mass and more spores. More stems, then stems branching off of stems, developed. Other now-favorable mutations allowed a little photosynthetic epidermal tissue on the stems to extend beyond the stem, conferring a competitive advantage in the sunlight-harnessing business. Ultimately, the tissues between stems met and merged, an evolutionary process much like the one that would produce webbing between the toes of the ancestors of ducks. Typical of the leaves that emerged from this process were the *pinnules* (little leaflets) of a fern frond.

The pinnules of a fern frond. Cell divisions along the edges of small green branchlets may have originally produced the web of leaf tissue.

Although the individual pinnules were small, they collectively gathered far more solar energy than bare stems, and the plants bearing them could grow tall. By the late Devonian era, the world's first forests—jungles, really—developed, covering the landscape in a dazzling diversity of green. Shrubby ferns grew in thickets on the forest floors. Other ferns grew into palm-tree shapes and sizes. The horsetail, or *Equisetum,* which grows in nodes like bamboo with a ring of needlelike leaves protruding at each node, rose ten stories tall. (The general morphology of their descendants remains unchanged, but extant species are only three to ten feet tall. These survivors, having withstood the test of many climate changes, are a gardener's nightmare, nearly impossible to fully uproot and unaffected by most herbicides.) *Lycophyta* with trunks six feet wide and 140 feet tall—about twice the diameter and height of a mature oak—dominated the landscape. (The only survivors of these giants are tiny club mosses.) Some species had stiltlike roots, a brilliant adaptation to the often swampy conditions of the time. At the end of the Devonian, *Archaeopteris,* the first tree with a trunk made of concentric sheaths of lignin and cellulose—that is, wood—emerged and thrived. (Tree ferns had trunks made of stems woven together; *Equisetum* trunks were hollow, like bamboo; and the giant lycophytes had a hard casing, but a spongy interior.) We would have found *Archaeopteris* odd-looking: At the top of its trunk, lateral branches emerged supporting fernlike fronds. About twenty-five feet tall, it had the most extensive root system yet evolved, much more massive than that of the taller lycophytes and tree ferns.

By the end of the Carboniferous era (360 to 300 million years ago), the Earth had been home to 150 million years of madly photosynthesizing vegetation. The continents were blanketed with greenery filling every vertical niche, from the soil surface covered with mosses to the canopies of lofty trees. Wafting oxygen as waste, plants had transformed the atmosphere. As much as 35 percent of the atmosphere was then oxygen, compared to today's measly 21 percent. That oxygen transformed the kingdom of animals that lived in this deep green world. Some of the descendants of the tiny fauna that had scuttled among the *Cooksonia* had evolved to gigantic proportions. Dragonflies with the wingspans of crows swooped through the air, three-foot-long millipedes and cockroaches the size of mice ran through the undergrowth, and two-foot-long water scorpions trolled the shallow waters. Amphibious newts were the size of crocodiles.

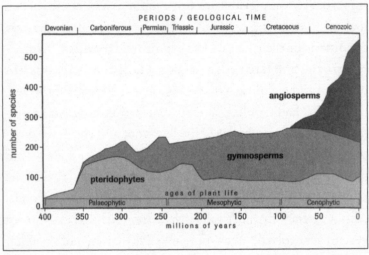

The earliest plants didn't need leaves because carbon dioxide was plentiful in the Devonian. During the Carboniferous, as soil levels increased, plants had access to more nutrients, and forests developed. During the Triassic, conifers—members of the gymnosperm division—thrived in the hot and dry climate. Although, with their narrow leaves, they were slow growing. Large-leafed, fast-growing angiosperms predominated during the Cretaceous.
Chart by permission of Kenrick and Davis and Niklas.

Every minute of those 150 million years, trunks, branches, roots, and leaves had been growing, dying, and rotting. Soils became a meter deep and more in places, enough volume for ever-larger root systems. More plants grew and died than could be decomposed. (Whether this was because bacteria had not yet fully mastered the breakdown of lignin-rich wood or because plant material fell into anoxic swamps where decomposers didn't live is unclear.) Over the eons, those plants were buried, compressed, and eventually became the coal— hence the designation of "Carboniferous"—we burn today. Practically overnight, considering that the continents had been essentially barren for 4.1 billion years before *Cooksonia* evolved, chloroplasts had transformed the planet. By steadily absorbing carbon dioxide, splitting water with solar power, and snapping together the carbon-oxygen-hydrogen mole- cules of sugars, they had created our familiar world.

nineteen

The Tenacity of Trees

Isn't it always so? The beautiful hickory died an untimely death while our neighbor's trees, a raggedy row of Leyland cypresses planted right on the property line, appear to be immortal.

In our close-in suburban neighborhood, most people's lots are well less than a quarter-acre. We accept that we will live with a view, only marginally obscured by shrubbery and backyard fences, into each other's yards and lives. I know, for example, when Doug across the street is sleeping in: His newspaper sleeps, too, wrapped in its plastic sleeve on his front steps. I knew before the Thompsons that sparrows were nesting in their eaves. Louise told me over the back fence that our dog had treed a raccoon one midnight. Bob across the street once called me to report that Alice, age seven at the time, was playing on our sloping porch roof. Our neighbors to the north, however, a physics professor with appropriately Einsteinian hair and his elegant French wife, have always been determined to have privacy, and more than twenty years ago, planted all four sides of their lot with Leyland cypress saplings.

If allowed to develop naturally, *Cupressocyparis leylandii* grow as much as fifty feet tall and in a dense, narrowly conical form, tapering from fifteen feet across at the base upward to a point. But because the professor planted his saplings a mere three feet apart, the trees have grown up to look nothing like that. Most of their lower limbs, too deeply shaded, have died, so the trees that border our property now resemble thirty-foot-tall fence posts with branches only in their upper third. These branches, desperately seeking unobstructed light, extend mainly to the north and south (the row runs east-west). During winter, wet snows inevitably prove too much for one or two individuals, and we wake to find them listing drunkenly over our roof. In the spring, our neighbor's yardman wires the fallen fellows to their sturdier comrades in order to reestablish the line. Nonetheless, although grotesquely pruned by snows and now laced together like a stockade fence, the Leylands survive year after year.

The trees pose some risk to our roof, but more to my spirit. Our house is close to the edge of our sliver of a lot, and the cypresses are a looming presence. One dismal winter day, aching for more light, I typed "how to kill Leyland cypress" into my browser's search box, and discovered that I am far from the first to harbor sanguinary feelings about them. Leylands have become known, according to the BBC Newsmagazine, as the "scourge of suburbia" and "a by-word for neighbourly bust-ups." In 2001, in the town of Talybont-on-Usk in Wales, one Llandis Burdon, age fifty-seven, was fatally shot in the course of a dispute over Leylands. The British government estimated in 2005 that there were as many as seventeen thousand unresolved disputes between neighbors over towering stands of these trees. In England, Part VIII of the Anti-Social Behavior Law, sometimes called the "Leylandii law," gives authorities the power to force a homeowner to lower the height of his hedge should a neighbor complain.

It appears from my Internet browsing that Leyland cypresses are not easy to kill. Simply chopping down the beasts doesn't slay them: Their stumps, hydralike, quickly generate multiple new sprouts. One Internet poster suggested coiling a soaker hose on the soil around their trunks and leaving the water on (for how long, he didn't say) to drown the roots. Several others suggested spreading a heavy dose of rock salt on the ground, killing them à la Carthage.

That cypresses would be hard to subdue makes sense. *Leylandii* is a member of *Coniferales,* a division of woody trees that emerged late in the Carboniferous to take advantage of the new, deep soil. Conifers have survived ever since, through hothouse and icehouse climates, moist and arid conditions, mountain upheavals, and several catastrophic asteroid collisions that killed off many less hardy species. Most of their primeval companions have long gone extinct or been reduced to tokens of their former selves, but conifers are represented today by some 630 species. Some of the world's most venerable specimens of trees are conifers, including the tallest living tree (a 380-foot Sequoia) and the oldest tree (a five-thousand-year-old bristlecone pine). Even older, in a sense, is a spruce growing in Sweden. The visible portion of the tree is only thirteen feet high, but its root system, which sends up a new shoot to replace the old one (after six hundred years or so when it dies), has been alive for more than 9,500 years.

Conifers are *gymnosperms* (JIM-no-sperms), a new group of plants that evolved in the late Carboniferous. Instead of reproducing via spores, gymnosperms had seeds. Spores are single cells. Seeds, on the other hand, contain embryos—open up a fertilized and ripe seed and you may see an embryonic root, shoot, and a leaf or two—inside a multicellular seedcoat. The seedcoat also encloses a stash of carbohydrates, the *endosperm,* that the embryo uses to grow before its leaves develop sufficiently to gather solar energy. But while the gymnosperms had larger, more complex seeds, they had to sacrifice more of their stored energy to produce them. Consequently, they couldn't afford to produce as many seeds as ferns produced spores, although any individual seed was more likely to thrive than any single spore.

At first, this trade-off was a viable, if not an overwhelmingly successful, strategy. The earliest conifer species were small, and some looked much like that common, often rather spindly houseplant, the Norfolk Island pine. They had similar needlelike leaves and sparse, resinous branches that extended in a whorl from a slender trunk, although their cones were more primitive. From the end of the Carboniferous and through the Permian (about 300 to 250 million years ago), a handful of conifer species persisted. Then, suddenly, at the end of the Permian, the *Coniferales* exploded in number and diversity, and came to dominate the Earth's flora.

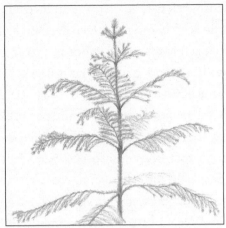

A Norfolk Island pine.

Why? First, a geologic accident that had been waiting to happen, happened. The planet's continents, always moving on tectonic plates, collided to form Pangaea, a supercontinent that stretched from one pole to the other. As a result, ocean circulation patterns were interrupted. Sea levels dropped as glaciers formed at both poles, and the shallow, warm continental shelves, once full of diverse marine life, became land. Around the equator, the many inland lagoons and swamps dried up entirely, replaced by sand dunes. Land in the mid-northern and mid-southern latitudes became temperate, and as is the case today, had seasons, including chilly winters.

Next, it seems that an asteroid six miles wide and the mass of Mount Everest slammed into the Earth off the coast of northwestern Australia, creating a 125-mile-wide crater and adding massive amounts of light-blocking dust to the air. Unable to photosynthesize in the haze, plants and algae died, thereby starving many of the marine animal, amphibian, reptile, and insect species that depended on them. Then, massive volcanic eruptions in Siberia spewed out lava that covered at least 1.5 million square miles, an area larger than Europe. Because the lava erupted through the largest coal basin in the world, large amounts of carbon dioxide and sulfur dioxide rushed into the atmosphere, trapping heat and causing dramatic warming. The trapped gases melted the permafrost and heated even the frigid seabeds, which in turn belched sequestered methane, an especially potent greenhouse gas.

The sequence of these events and their relative importance is debated, but there is no question about the outcome. In whatever form and order death arrived—gas poisoning, freezing, burning, starvation, suffocation, and dehydration—we know that life was decimated within a two-million-year period, a blink of a geologic eye. The catastrophic confluence of events is called the end-Permian extinction or, colloquially, the Great Dying. Ninety-six percent of all marine species went extinct; at least 90 percent of all life on Earth died. Earth at the end of the Permian and the beginning of the Triassic (250 million years ago) was, in most places, hot and dry, with shockingly reduced biodiversity.

Only one family of conifers survived the Great Dying, but that family, the *Voltziaceae,* was well positioned to thrive. Its needles were perfectly suited for the hot and dry climate. They had a small surface area and were heavily coated in a waxy substance that resisted desiccation. Their stomata were sunken into the needle's surface to further limit evaporation. Because the concentration of carbon dioxide in the atmosphere had increased, they were nonetheless able to access sufficient amounts of the gas. The conifers thrived, radiating into multiple families and at least twenty thousand species. Many were tall (some up to two hundred feet tall, three times the height of an oak) with short branches radiating from top to bottom, and looked like gigantic bottlebrushes; others, like modern sequoias, were tall with thick trunks sporting stiff branches only at their summits. Some evolved to cope with high salt concentrations as mangroves do now; others, in order to survive on flood plains where any individual's life was short, set cones in a season or two. A few species had the Christmas tree shapes reminiscent of our spruces and firs. By about 200 million years ago, conifers dominated the land, comprising about 50 percent of the world's plant species.

Then, in the heyday of dinosaurs about 140 million years ago, something new under the sun appeared. The first plants with flowers—the angiosperms—evolved in the tropics, diversified, and expanded their domain rapidly. The secret of their success was their leaves. Broad and flat, they had a greater surface area that gathered more sunlight than the narrow-bladed conifers. The level of carbon dioxide in the air had fallen, and these new leaves could access more of it. Making more sugars, they could grow faster, especially as seedlings and saplings, and they shaded out the slower growing conifers. Conifers eventually ceded great tracts of territory, surviving primarily in those regions too cold for most angiosperms or in poor soils that couldn't support angiosperms' high demand for nutrients. Today, 75 percent of all species are angiosperms, conifers are 15 percent, and pteridophytes—ferns, horsetails, and club mosses—comprise the remaining 10 percent. The largest concentrations of conifers are in mountainous areas, the far latitudes, and in rocky, clayey, or acidic soils.

Among the remaining conifer species are two natives of North America: the Monterey cypress and the Nootka cypress. The former grows wild only around Monterey and Carmel in California, along the cool and rocky coast. It is that iconic wind-sculpted and twisted form often photographed half obscured by fog. The few specimens, some two thousand years old, are all that are left of what once was an extensive forest. The Nootka has a traditional pyramidal profile and distinctive foliage that droops in sprays from its branches. It grows in high altitudes along the coast from northernmost California to southern Alaska.

On the left, a Monterey cypress; on the right, a Nootka cypress; in the middle, a Leyland cypress, a sterile hybrid of the two.

These two august species are the parents of the Leyland cypress, a sterile cross between the two. The pair never would have met in their native habitats: The southernmost Nootka grows four hundred miles north of the Monterey range. But in the mid-1800s, Christopher Leyland, a banker in Liverpool, gave his nephew John Naylor an estate in southern Wales as a wedding present. At great expense, Naylor renovated the house and hired a landscape architect to lay out gardens.

The architect installed a variety of exotic trees, including the Monterey and Nootka, which he planted in close proximity. In 1888, a hybrid seed sprouted and grew. The next year, Naylor died, and his son Christopher inherited the estate and changed his surname to Leyland. Christopher Leyland took six seedlings to his own property in Shropshire, where they rapidly grew into large trees, *Cupressocyparis leylandii*. Because Leyland cypresses are sterile, the specimens that straggle along at the edge of our yard are the direct descendants, by cuttings, of Christopher Leyland's trees.

Of course, I never attempted to kill my neighbors' Leylands; a premeditated murder is beyond me. Besides, large trees evoke in me—and myth and folklore tell me I am not alone—a deep-seated reverence. In any case, I know that *leylandii* are not long-lived, and the ones next door are approaching their natural end. In fact, my neighbor has anticipated their demise, and just inside the row of trees he has planted a long line of shrubs. The shrubs are *Nandina domestica,* a flowering species that should grow into a well-behaved, six-foot hedge. Which means that right here in Maryland I will be witnessing a reenactment of ancient history, as angiosperms replace the conifers.

twenty

Amazing Grass

Some midwinter day when you're in the grocery store, pick up a few boxes of cherry tomatoes and read the labels to see where they were grown. Most come from Mexico. That makes sense: warm climate, long hours of sunlight. Others are from Canada, grown in greenhouses. The strange thing is that both boxes are about the same price. How can a Canadian grower who must pay for heat compete with the Mexican grower who gets all his therms for free? In the summer of 2011, I set out to find the answer at Pyramid Farms in Leamington, Ontario, where owner Dean Tiessen has thirty-seven acres of vegetables under glass roofs. As soon as I pull into the farm's office, having driven about an hour southeast from Detroit, Dean bounds out to greet me. He is a fit and handsome man in his mid-forties with a straight-up shock of dark hair.

If anyone has farming in his blood, Dean does. His fore-bears were Dutch Mennonite farmers invited by Catherine the Great in the 1760s to settle and modernize farming in southern Ukraine. There they stayed, farming lucratively generation after generation, until the communist revolution in 1917. Dispossessed by collectivization, his grandparents fled to Canada and settled in Leamington, where, on one and a half acres, they grew seedless cucumbers and tomatoes in greenhouses. The farm passed to Dean's father in the 1950s, and about ten years ago Dean, his brother, and two cousins took over. They transformed a small operation that sold into the local market into a business that supports three families, employs more than a hundred people, and sells across North America. Pyramid Farms now competes in a highly price-sensitive, global market.

So how does a Canadian succeed? Dean slides open a greenhouse door to show me. Forget tomato bushes. I am looking into an eight-foot-tall solid wall of tomato vines that extends the sixty-foot length of the greenhouse. It is densely hung with tomatoes, the largest, reddest ones toward the bottom, little green ones at the top. I peer through the wall, and see another one just a few feet behind this one. Dean tells me there are about a hundred tomato walls—he calls them rows, but that doesn't do justice to their bulk—in each greenhouse.

This is tomato growing at its most intensive and efficient. We look at the base of one wall. Forget soil. Two tomato vines, thick as ropes, emerge every foot or so from a foam block set in a narrow trough in the concrete floor. A black umbilical cord of water and nutrients runs into each block. Far above, a horizontal wire runs the length of the greenhouse just below the ridgeline. Spools of string hang down from the wire every few feet. Each vine is assigned its own string and has been trained to grow up along it. Every two weeks, the spools move farther along the horizontal wire and unwind about two feet of string. Every week, a worker on a lift twirls a newly grown length of vine up the bare string toward the overhead wire. The tip of each vine grows farther and farther from its base. Eventually the tips will be sixty feet from their roots. The lengthening of the strings effectively lowers the older portions of the vines, so that great ropes of parallel, leafy, tomato-filled vines slope very gradually from floor to ceiling.

Eventually these vines will grow to sixty feet in length.

At the top of the vines, new leaves and a cluster of little yellow flowers emerge. I run my eyes down a vine and count eight clusters of tomatoes. The tomatoes in the topmost cluster are small and green, and each successively lower group is in a greater state of ripeness. By the time the tomatoes are perfectly ripe and bulging, they're about knee height. The low drone I hear in the greenhouse comes from air circulation fans, but it could be the thrum of tomatoes growing at full throttle. Every leaf in here is green and healthy; every fruit is blemish-free.

Dean tells me he can now grow as many pounds of tomatoes in one acre indoors as his Mexican counterparts can grow outdoors in forty-seven. In the last ten years, he has tripled production per acre, thanks to improvements in greenhouse technology and the breeding process. He also has refined his crop selection, choosing to grow only specialty tomatoes—including twenty-six varieties of heirlooms—that have higher profit margins. The only variable he can't improve is the Canadian climate and his concomitant need for fuel. About 40 percent of his cost of production is energy, and he feels the pain of every penny increase. More than any other worry—competition, blight and bugs, labor costs—it is the volatility of energy prices that keeps him up at night.

"My father burned coal to heat the greenhouses until 1967," Dean tells me. "Then oil refining became a business in the Port Huron area, and he burned 'bunker oil,' the thick oil left at the end of the refining process. By the time I got involved in the business, the infrastructure for natural gas had arrived, and I kept switching between gas and coal, going back to coal when it was cheap. Then, in 2002, all fuels skyrocketed. Our heating costs went from thirty thousand dollars an acre to one hundred thousand dollars in one season.

"The situation was dire, and I looked around for any alternative, and ended up moving into wood. There's no forest around here, but I found construction debris that otherwise would have ended up in a landfill. Every time I saw someone tearing down a building, I was there asking if I could haul away the lumber. For a while, it was fantastic: I saw my energy costs drop to twenty thousand dollars an acre. But soon everyone was going after the stuff, and build-ers stopped giving it away. It became a commodity. Then it got scarce, and the price went way up. Fortunately, by that time coal had become cheap again. Right now I'm burning natural gas."

About five years ago, Dean explains, his inability to get "a line of sight" on future fuel costs and his experi-ence burning lumber inspired him to look into biomass for energy. If he could grow his own fuel, he might fix his long-term energy costs and sleep better. Maybe, with fixed energy costs, he could offer longer-term sales contracts for his tomatoes, which would attract buyers. And he figured that if a cap-and-trade system for carbon emissions ever develops, as it has for sulfur emissions, he could sell his carbon credits.

In 2006 he took a tour of European biomass farms. The English and the Germans had more varied experience with growing biomass than North Americans, who were focused on fermenting corn into ethanol to supplement gasoline. He stopped in on farmers cultivating willows and poplars, Japanese knotweed, switchgrass, and miscanthus. Knotweed turned out to be an invasive species in the United States, and therefore a nonstarter. Willows and poplars, while fast growing for trees, nonetheless take thirty years to get to harvest. In the interim, the crop could be devastated by disease, insects, or fire. Switchgrass, a perennial grass native to the North American plains, was an interesting possibility. But even more attractive was miscanthus, a perennial grass native to Asia and Africa. In Germany, researchers were having success with a hybrid called *Miscanthus giganteus,* a variety that grows as tall as twelve feet, and produces at least twice as much biomass per acre as switchgrass.

In Dean's eyes, the crop had additional attractions. Not only is it a perennial; it has proved to be particularly persistent. He saw experimental plots in Germany that had been growing for two decades. In Japan, where miscanthus has been cultivated for centuries as roof thatch, some stands are two hundred years old. Because *giganteus* is a sterile hybrid, it couldn't go to seed and escape his farm and invade his neighbors' fields. Nor would it, like kudzu, colonize by creeping: After twenty years, Danish experimental stands have expanded by only a few feet. If the crop didn't work out, it would be easy to uproot. Pests have no interest in its tough leaves, and after the first year, it grows so tall so quickly, it shades out its weedy competitors. Miscanthus can remain in the field, straw-colored and sere, until late fall or even spring, when idle harvesting and baling machines can take it down. The longer it stands in cold weather, the drier—and the better for burning—it gets. Bales of miscanthus can be left in the field for months without degrading, so there would be no storage costs.

Unlike the corn grown for ethanol in North America, miscanthus grows well on marginal land too steep, too sandy, or too low in fertility to grow row crops. In the late fall, as it stops photosynthesizing, it sends the nutrients in its stalks and leaves back underground, which means a field of miscanthus needs few or no costly fertilizers. And *giganteus* is unpatented and freely available. The only downside seemed to be that no one had yet figured out how to plant it efficiently, so initial planting costs would be high, but Dean figured he could overcome this. On his return from Europe, Stephen Long, professor at the University of Illinois and one of the world's leading miscanthus researchers, gave him five rhizomes to try in Ontario.

Dean takes me outside to see one of the miscanthus fields descended from those rhizomes. From a distance, the field reminds me of the blocky mesas that rise abruptly out of the landscape in the American Southwest, except, instead of ochers and beiges, here the hues are emerald and dark green. As we approach, the mesa reveals itself to be a dense mass of vertical canes well clothed in bladelike leaves. The whole assemblage sways and shivers in the morning's brisk breeze. Dean urges me to stay back a minute to take a photograph, and he strides ahead to position himself at the front edge of the field. With Dean in the photo, it is clear that the plants are already nearly twice his six-foot height, and it is only mid-July. *Giganteus,* indeed.

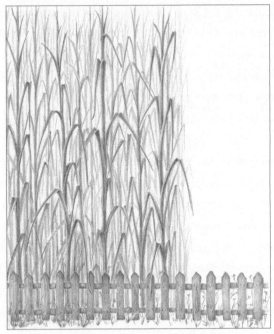

A stand of *Miscanthus giganteus.*

When Dean and I stand at the edge of the field to look more closely at the plants, I can see that the leaves, dark green with a thin, white stripe down the midline, emerge at every joint of a segmented cane. With two hands, Dean grabs a cane near its base and, tugging hard, pulls it out of the ground, along with its subterranean anchor. He is careful not to grab the leaves, which are covered in microscopic silica, a deterrent to insects and small herbivores. (He recently got in trouble with his wife, he tells me sheepishly, for letting their two youngest sons play hide-and-seek in a miscanthus field. When his wife put the boys in a bath with Epsom salts that evening, they shrieked with the sting from invisible cuts.)

He snaps off the cane and hands me what looks like a piece of thick, gnarly root about eight inches long. It's not really a root, he explains, but a rhizome, a stem that grows horizontally underground. The rhizome is segmented, and each segment or node has a large and stubby bud. The buds are capable of producing either a new rhizome, a fibrous root, or a new cane, depending on the hormonal signal they receive. Over time, many of the buds will become new canes.

Although it is the tough canes that are harvested for biofuel, it is miscanthus's leaves that are key to its success. Grasses are newcomers to the planet's flora, coming on the scene just as the dinosaurs vanished 65 million years ago at the end of the Cretaceous. No one knows for sure why they evolved so late in the geologic day, but climate change likely played a role. At the time, the higher latitudes in the continents' interiors were becoming more arid and fires more frequent. Grasses are well adapted to fire because their growing tips are at or even below ground level where, sheltered from flames, they can regenerate that very season. Trees, on the other hand, are either killed outright or take years to recover.

The evolution of large mammals in the post-Cretaceous era also helped the grasses evolve. By 55 million years ago, grazing hoofed animals, the ancestors of modern horses, antelopes, cattle, and camels, were clipping the tender tips of shrubby plants and small trees, stunting or killing them and opening more territory for grasses. Grazing animals might munch grasses to the ground, but when the herd moved on, the grasses sprang up again. By ten million years ago, temperate regions were covered in vast grasslands much like the modern prairies of North America, steppes of Eurasia, pampas of South America, and veldt of southern Africa.

About that time, in hotter, drier environments closer to the equator, a new class of grasses, what you might call supergrasses, proliferated. These species are today's critical food crops of sugarcane, corn, millet, and sorghum, as well as bamboo and (ta-da!) miscanthus. The key to their success was their reinvention of photosynthesis.

Most plants and the older grasses photosynthesize in a "C3" fashion, and as successful as they were and are, they have a physiological weakness. One of the steps in fixing atmospheric carbon dioxide into sugars involves the enzyme RuBisCO. But, RuBisCO regularly makes a mistake and fixes oxygen instead of carbon dioxide. The plant then needs to shed that oxygen, and in doing so, loses some carbon it recently trapped. In hot weather, the problem gets worse. When plants need to conserve water—that is, when they are losing more water through evaporation from their leaves than their roots can replace—their stomata automatically close. When the stomata close, waste oxygen can't escape and accumulates—and is fixed by RuBisCO—in the leaves. In hot conditions, C3 plants make fewer sugars.

Certain grasses in the tropics evolved a couple of anatomical and physiological tricks that allowed them to get around this inefficiency. These "C4" plants developed a new biochemical pathway (only discovered by scientists in 1966) that starts by putting carbon into a four-carbon, instead of a three-carbon, compound (hence "C4" and "C3"). That compound is pumped into bundle-sheath cells, the cells that surround leaf veins. In C4 plants, the bundle-sheath cells themselves photosynthesize, while in C3 plants they typically do not. More important, in the C4s these cells are able to concentrate carbon dioxide at a higher level than is in the atmosphere. RuBisCO therefore contacts and interacts with more carbon dioxide molecules than it otherwise would, so more sugars are created. Because the bundle-sheath cells are impermeable to oxygen, less wasteful oxygen fixation occurs.

In sum, C4 plants learned how to make more hay—about 40 percent more—while the sun shines. That is why although they represent only 1 percent of the world's plant species, they represent 20 percent of the Earth's vegetation coverage and produce about 30 percent of terrestrial carbon. That is also why fourteen out of the eighteen of the world's worst weeds (we're looking at you, crabgrass and pigweed) are C4 species. And that is why miscanthus is a prime candidate for making biofuel. Moreover, it turns out that among C4s, miscanthus is especially good at accumulating carbon, even in a climate as untropical as Ontario. In the spring, miscanthus sends up new shoots from its rhizomes several weeks earlier than corn, which has to develop from seed, and even switchgrass. Its leaves continue to photosynthesize weeks after those two have called it a season. I am not surprised to learn, looking at the impenetrable mass of leaves in front of me, that miscanthus has more leaves with a larger collective surface than other C4s.

By the spring of 2008, using a combination of microscopic tissue culture and by manually dividing and redividing rhizomes, Dean turned Dr. Long's five rhizomes into thousands, and planted them on sixty acres. By the end of one growing season Dean had ten times the number of plants that the university had. By 2014, Dean expects his tomato farm will be energy self-sufficient.

The psychological benefits of energy independence are considerable, but does growing miscanthus make business sense? Dean has four hundred marginally productive acres of his own in miscanthus, and he contracts with neighboring farmers to grow the rest. Dean figures that the cost of a gigajoule (a billion joules) of miscanthus-made energy is $4.20. Natural gas, which is cheap at the moment, costs about $6.50 per gigajoule. It doesn't take a special furnace to burn the biofuel. At least on paper, miscanthus fuel is a winner.

The environment comes out ahead, too. Burning coal, natural gas, and oil that has been sequestered underground since the Carboniferous era releases climate-warming carbon dioxide into the atmosphere. While burning miscanthus in the winter adds carbon dioxide to the atmosphere, miscanthus took that carbon dioxide from the atmosphere in the spring and summer as it grew. The only fuel Dean uses in producing miscanthus energy is a little diesel for harvesting and baling the crop. In three years, he expects his farm will be nearly carbon neutral, which in a world where carbon dioxide levels are rising at an unprecedented rate is a significant accomplishment.

There are other environmental benefits, as well. Over the course of a summer, as the canes grow, the older, lower leaves fall off. They decay gradually, and will eventually return carbon and nutrients to the soil, but until then, the thick leaf litter provides good cover for small wildlife. The fallen leaves also slow evaporation from the soil. The underground portion of a miscanthus plant is equal in mass to the aboveground portion, which means it has a substantial impact on the subterranean ecology. The perennial roots of miscanthus reach deep into the ground, where they aerate the soil, leak nutrients to worms and insects, and add organic material to create a rich subterranean ecosystem. In terms of environmental impact, it is a crop much like perennial wheat, the holy grail of the Land Institute. Even better, it grows where no self-respecting wheat would grow.

Denmark, Spain, Italy, Hungary, France, and Germany have started multiple research and commercialization projects, and the European Union projects that 12 percent of its energy will come from miscanthus by 2050. In the United States, the USDA is supporting projects that are expected to have 100,000 acres in miscanthus. BP recently invested $500 million for miscanthus research at the University of Illinois.

I am fascinated by the prospect of new biofuels in general, and intrigued by miscanthus. When Dean tells me it would probably take about an acre and a half of miscanthus to heat a typical house in Ontario, I mentally add a good stand of miscanthus to my dream house, the one in the country where I won't have to wonder whether a neighbor's *leylandii* will fall on my roof. Miscanthus can't fill all our energy needs. You can't put miscanthus directly in your gas tank. But thanks to a million years of evolutionary fine-tuning, each leaf is a marvel of a machine for turning sunlight into stored energy.

PART IV

Flowers

twenty-one

Sex in the Garden

It was about noon on Saturday nearly thirty years ago, and Amy, my best friend since childhood, and her fiancé, John, who had just graduated from medical school the previous day, were getting married at four o'clock. Tomorrow, they would be leaving Washington, D.C., driving a U-Haul to Michigan, where John would start his residency. A congenial cabal of John's and Amy's sisters and friends—some busy in the kitchen, others packing boxes in the hallway—were helping to make this happen. The living room furniture was still in place in their garden apartment, but the bedroom was impassable, crowded with bicycles and cross-country skis, stacks of pine shelving, green milk crates, garbage bags filled to rotundity, suitcases, backpacks, framed posters blanketed in bubble wrap, and a craggy mountain of cardboard boxes, all neatly labeled in purple marker in Amy's cryptic, slanting demi-script. Meanwhile, staff from the party rental company were erecting a tent in the backyard of the building and setting up tables and chairs for seventy-five guests. There would be a potluck dinner, and salad bowls and Pyrex serving dishes and bread baskets—more arriving by the moment—were lined up on the counter of the small kitchen. The cake was in a cooler; the champagne and wine were on ice in the fridge, which was still covered with the artwork of Amy's third graders. The only thing that was missing was the flowers.

There was a knock at the front door, and a grinning taxi driver presented Amy with a stack of long, shiny white boxes tied with white satin ribbons. The flowers had arrived, the generous gift of John's stepmother, Betty, who lived in Honolulu and had brought them with her on the plane. The hotel had agreed to keep the boxes refrigerated. We—the sisters and friends who happened to be in the apartment—gathered in the living room to watch Amy open them. She had no idea what to expect; Betty had solicited no suggestions nor offered any choices. Cue the oohs and aahs: Inside the first boxes were sprays of white orchids, starbursts with thin lines of lavender running down their narrow petals. Other boxes held lovely, traditional *Phalaenopsis* with rounded, deep purple petals. Delicate yellow blossoms emerged from another set. Then Amy opened one of the three last boxes. Inside were long green stems, each topped with what looked like a red, heart-shaped, plastic plate. At the center of the flower, a fleshy, pink column, longer than my middle finger and about as thick, jutted out at a ninety-degree angle. We stared, and then—pardon the cliché—exploded with laughter. The last two boxes held more of the same.

Someone said, "They're anthurium."

Amy said, "Well, whatever they are, I'm not having them at my wedding," and no one disagreed.

I went to the kitchen to find vases and fill them with water, and by the time I returned to the living room, only the orchids remained to be arranged. Mixed together, they were even more beautiful than they were segregated, a nice metaphor for the coming event.

It was as high-spirited a wedding as I'd ever been to (the bride wore a tulip-print dress with an eyelet ruffle at the hem and the groom was shod in green tennis shoes) and, happily, did not mean the end of a friendship, as sometimes a marriage and a move do. Recently, on the annual midwinter escape that Amy and I take to Puerto Rico, I mentioned that I was writing about anthurium. Whatever happened, I asked, to those flowers that Betty had brought to her wedding? She laughed, and was surprised she'd never told me the story. Her father had booked a post-wedding suite for them at the Tabard Inn, a small and dignified hotel in downtown Washington. The Tabard is a combination of three old townhouses dating to the 1880s and the rooms are decorated with hand-hooked rugs, porcelain table lamps, marble-topped dressers, and the like. When she and John walked into the room late that night, there, on the white lace coverlet on the mahogany four-poster bed, amid a shower of glittery red confetti, were the anthurium. Her sisters had spirited them away that afternoon, begged a key from the front desk, and done a little prenuptial decorating. It was a funny moment, Amy said, but even today, she finds the flowers something of a salacious joke and can't see how anybody could like them.

I can. I love anthurium.

I do all my gardening indoors, in the glass-filtered light of a conservatory. No outdoor gardening with dog's-breath heat and humidity, murderous mosquitoes, and horrible hundred-legged beasties for me. My soil comes, bugless, in plastic bags. Water—into which I carefully measure teaspoons of fertilizer—comes from the spout of a watering can. None of my potted plants are native. All hail from well south of here, and that is my challenge: nurturing a tropical paradise where none has a right to be. Moreover, not only do I live at the wrong latitude, but my conservatory roof faces north. As a result, profusely flowering plants are difficult to sustain, and my domestic jungle is primarily green, rich in textures and shapes but not in color. Slender reeds of *Papyrus* and the fronds of a Majesty palm reach toward the ceiling. Asparagus ferns hang from pots in frilly abundance. In the western window lives a jade plant with its shiny, plump leaves, a dwarf banana (which, like the Bird-of-Paradise, has never flowered), a classic grandfather cactus, a "pencil cactus" that looks like a tangle of skinny green sticks, and an agave so dangerously spiky, I wear oven mitts when I need to rotate it. A few species offer a bit of leaf color: *Maranta* (prayer plants), for example, sport pink-veined leaves and *Dracaena marginata* "Tricolor" have stiletto-like leaves striped from stem to tip in pink, cream, and green. My citrus trees produce tiny white flowers, thanks to the grow lights, but only for a week.

For real color—and I need it when night falls early and the dark windows seem to suck the heat out of my body and the high spirits out of my heart—I rely on anthurium. There are about a thousand species of the *Anthurium* genus, and I have varieties whose *spathes* (the proper name for what looks like a single, giant petal) come in every color on the L'Oreal nail polish display: stoplight red, traffic-cone orange, deep peach, sultry magenta, porn pink (oops, that's not on the chart), and every hot hue in between. The *spadix,* that phallic central column, is flashy, too, and can be bicolored in red and pink or gold and cream. Not only are anthurium content with northern light; they retain their vivid colors for weeks and often months.

Amy is not the only one who does not appreciate anthurium. Ted is not a fan. When the plants are all in bloom—and I usually have about a dozen of them—he says the conservatory looks like a bordello with the customers caught in flagrante delicto. I have seen how they make guests uncomfortable. Having a dinner party in the conservatory among anthurium is not like picnicking in a field of sweet meadow flowers. Not unless the Marquis de Sade happens to be lounging *en déshabillé* on your picnic blanket.

But, of course, even the white daisies, pink clover, ivory Queen Anne's lace, and lavender lupines in the meadow are all about sex. Their pretty petals and delicate scents are the come-hither signal to insects, an invitation to sample the sweet rewards inside. On the way to sip the nectar tucked away at the base of the flower, an insect—the only real innocent on the scene—picks up pollen that will fall off inside the next flower it visits. Ironically, when it comes to anthurium, the anatomical part that makes us blush is just a sturdy stem that hosts the plant's flowers. Those flowers are so minute that you need a magnifying glass to see them. Except to human eyes, the anthurium is actually a model of floral modesty.

Anthurium do not grow in Europe, although other spadix-sporting *aroids* (as members of the Arum genus are called) do. The Cuckoo Pint (*Arum maculatum*) grows in temperate northern Europe. *Pint* is a shortening of *pintle,* a medieval word for penis. *Dracunculus vulgaris,* which grows in the eastern Mediterranean, features a two-foot-long dark purple spadix. Nonetheless, these species inspired no medieval or Renaissance Europeans to suppose that flowers had anything to do with procreation. (Theophrastus's report of North Africans shaking the dust of male date palms over female palms had slipped from common knowledge.) After the fall of Rome in the fifth century, the keepers of knowledge, both sacred and profane, became the clergy, and no group was less likely to interest itself in the sexuality of plants. By the medieval era, as the veneration of Mary and virginity grew, writers and artists linked her with a host of flowers. Marigolds, roses, rose of Sharon, violets, irises, orange tree blossoms, and many other flowers decorated the background of drawings and paintings that featured her. A sacred space featuring a statue of the Virgin surrounded by these flowers was known as a "Mary Garden." The white lily became Mary's signature flower in the twelfth century when the angel Gabriel was first pictured offering them to her at the Annunciation. Naturally, no one living in the era could have imagined that flowers, those inspiring reminders from God of the goodness of chastity, are actually an invitation to an orgy.

Nehemiah Grew was the first European since the classical era to posit plant sex. The stamen, he suggested in *The Anatomy of Plants* (1682), just might be the male flower's generative organ. It was possible that the anther served as "a small penis" and that the "small Particles" produced by the anthers were "Vegetable Sperme." When the particles fell on the "Seed-Case or Womb" of a female flower, he wrote, they impart a "Prolifick Virtue." This was astute speculation by a careful observer of plants, but it was conjecture only, and Grew did not attempt any experiment. John Ray, the notable English naturalist, seconded his countryman's notion, but Marcello Malpighi, equally authoritative, disagreed. He thought seeds develop the way flower buds do. Buds don't need fertilization to develop, so why should seeds? Pollen isn't sperm, he wrote, it is excrement.

Grew's proposal that flowers have genitalia generally went unremarked. It was, after all, only a thought and, for most people, a profoundly unsettling one. Rudolf Jacob Camerer, the director of the Botanic Garden at Tübingen, was one of the few who did take notice. Camerer was a generation younger than Grew and Malpighi. He started his investigation of flowers around 1690 by looking at the development of the embryos inside the seeds of bean and pea plants.

Both these species (and 90 percent of all flowering plants) have hermaphrodite flowers, meaning each flower on a plant has both male and female sexual organs, male stamens and female carpels or pistils. (To remember that stamens are the male organs, think of the "men" in *stamen*. Carpels and pistils are synonymous for my purpose here.) The stamens are composed of threadlike *filaments* topped by pollen-producing *anthers*. The carpels are composed of, from bottom to top, a rounded ovary with *ovules* inside, a slender *style,* and a flattened crown called the *stigma.* Looking through a microscope at the flower of a pea plant, Camerer realized that its ovules are initially filled with clear liquid. After pollination, the volume of liquid decreases in the ovule, and a small green point or globule appears. The globule becomes an embryo with two tiny leaflets (its *cotyledons*—pronounced ka-tull-EE-duns) and a tiny root and shoot. Camerer realized there had to be a relationship among these events. The flower's pollen, he hypothesized, fertilizes its ovules, which become the new embryos.

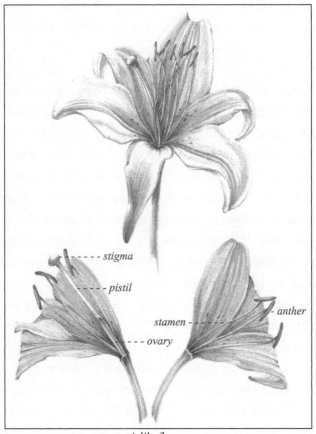

A lily flower.

Camerer ran a series of experiments with mulberry trees and dog's mercury, a shade-loving plant that carpets European forests, to test his hypothesis. He chose these species because they are among the few in northern Europe that are *dioecious* (di-EE-shuss), meaning the male flowers grow on one plant and the female flowers grow on another, and so, he reasoned, he could easily eliminate all male plants in a test plot. When he did so, the seeds of his female plants failed to ripen. He then ran similar experiments on *monoecious* (mon-oh-EE-shuss) plants, including maize, that have separate male and female flowers on the same plant. The two types of experiments proved to his satisfaction that if pollen doesn't fall on stigmas, either no seeds form or the seeds are sterile. There is, he concluded, no virgin birth in the plant world. Flowers have sex.

No lightning bolt seared him for this heresy, and in 1694, he wrote up his results in a letter to a colleague that was printed in *Transactions of the Tübingen Academy*. As A. G. Morton noted: "Few epoch-making papers can have had smaller circulations than this famous *De sextu plantarum epistola*." Still, how could this data, no matter how obscurely announced, go unheralded?

For one, there was some ambiguity in Camerer's results. Sometimes a few fertile seeds resulted from his experiments with maize and hemp. (In the case of maize, wild pollen may have drifted into the trial patch and fertilized the seed. As for hemp, it has a quirky sexuality: Generally dioecious, its female plants occasionally produce male or hermaphroditic flowers that can fertilize its female flowers.) There was also the puzzle of club mosses and horsetails. They appeared to have anthers but no carpels, so how could anthers be sexual organs? (What observers thought were anthers were actually *sporophytes,* organs that release spores.) The bigger problem was that Camerer's proposition that flowers are sexual was repulsive in the seventeenth century. Besides, one of the world's most influential botanists, Joseph Pitton de Tournefort, at the world's premier institute of botanical research, the Jardin du Roi in Paris, was unequivocal on the subject: Anthers were excretory organs. For more than twenty years, little further was written about flowers and sex.

Then, on June 10, 1717, at six o'clock in the morning a sub-demonstrator of plants at the famed Jardin, the forty-eight-year-old Sébastien Vaillant, stood up and delivered "the talk."

Vaillant was a substitute lecturer for Professor Antoine de Jussieu. Jussieu was a medical doctor, university-educated as all French physicians were, and the son of a well-to-do master apothecary, an eminence at one of the most prominent of the French guilds. The great Tournefort had died in 1708, and Jussieu, although only twenty-two years old, had succeeded to his professorship in botany. On the occasion of the inauguration of a new garden, he had been scheduled to deliver a lecture, but at the last minute had to travel to Spain. Vaillant, an older but more junior man at the institution, was chosen to speak in his stead.

Vaillant was an anomaly as a professional at the Jardin. His colleagues were all men of the nobility or the haute bourgeoisie, but he was the fourth child of a tradesman in northern France. When he was six, his father had boarded him with a priest who had given him religious instruction and taught him to read and write in French and Latin. Sébastien proved an excellent student and was musical as well, and his father scraped together the funds for harpsichord and then organ lessons, hoping his talented son might make a living as a church organist. At the age of eleven, the boy became the organist both at the cathedral in Pontoise and, in exchange for room and board, for the nuns at a nearby convent.

The nuns allowed their little musician to trail along with them when they worked at a nearby hospital, and Sébastien often slipped into the surgical theaters to watch the operations. Surgery fascinated him, and the surgeons, taken with his interest, loaned him books and—in an early and remarkable example of hands-on science at the elementary school level—gave him human body parts to dissect in his own room at night. Soon, he had taught himself enough to be accepted as an apprentice. He needed no academic training: Surgeons, affiliated with those other knife-wielding specialists, the barbers, were considered tradesmen. At twenty-one, with his apprenticeship completed, Vaillant moved to the suburbs of Paris to find work.

The move gave him the chance to attend public lectures at the Jardin du Roi in medicine and botany, those intertwined subjects. Here, he found his calling. He met Tournefort, and joined the other students who attended the renowned professor on his weekly field trips outside Paris. Vaillant took detailed field notes and wrote up descriptions based on his observations and, given that he was so handy with a knife, his meticulous dissections. The director of the Jardin, the noble-born Dr. Guy-Crescent Fagon, was impressed by his work and hired Vaillant to be his secretary, as well as the keeper of the Jardin's herbarium, or dried plant collection. When Vaillant was thirty-nine and an acknowledged master of plant anatomy and taxonomy, Fagon also appointed him as director of plant culture (in which capacity he introduced the first greenhouses to the Jardin) and "sub-demonstrator of plants," a position that entailed explaining the cultivation of plants and their uses to medical students and other garden visitors. The multiple appointments were necessary to provide Vaillant a living wage since the salaries of these positions were honoraria, appropriate for financially independent appointees. Of course, a professorship was out of the question for a mere surgeon.

After Jussieu departed for Spain, Vaillant posted the name of his upcoming lecture: "Discours sur la Sexualité des Plantes," which meant that the amphitheater on the morning of June 10 was packed. Flowers, Vaillant first declared, were the most important parts of plants. Stamens, which the celebrated Tournefort regarded as the lowest and most vile organs of plants, are actually "their most noble." Then he got down and dirty. Stamens are responsible, as in male animals, for the reproduction of the species. Their anthers are equivalent to testicles. In dioecious plants, he said, where the male flowers are at a distance from the females, "the tension or swelling of the male organs occurs so suddenly that the lobes of the bud are forced open with surprising rapidity. These male organs, seeking only to satisfy their violent transports, upon finding themselves freed, produce an abrupt general discharge, a swirl of dust, spreading fecundity everywhere. [Then] they find themselves exhausted." When a flower contains both male stamens and female pistils, "the stamens need not act with such haste or vigor, and . . . it may be presumed that the slower their actions are, the longer will be the duration of their innocent pleasures." All this, he said, could easily be observed in *Parietaria,* for example, in the mornings, but if one should happen to miss the action, there was an alternative. Provided "the plants, so to speak, have reached competent age," they could be spurred into action with the tip of a pin. In either case, the dust flies toward the female organs, where its "volatile spirit" travels down the solid style to fertilize the egg in the female's "belly." And so he continued in a scandalously anthropomorphic fashion.

The medical students loved the lecture: Not only was it prurient, but the lecturer had poked a middle finger in the collective eye of the stuffy establishment, always a thrill for the younger generation. Vaillant clearly enjoyed his subversive moment on the stage. He had long felt that Tournefort's major opus on classification, a new edition of which Jussieu was in the midst of preparing, was inadequate, but he had been unable to convince his superior to make any changes. Never had he doubted that he was the equal or better of Jussieu, despite the professor's elevated social and professional status. Surely it was a great pleasure to speak his mind and oppose conventional wisdom. I imagine he felt as liberated as those exploding anthers that morning.

Jussieu, on his return, was outraged. The lecture was an affront to the authority of the Jardin; the students' demands for more lectures from Vaillant were personally insulting. The French Academy of Sciences, shocked at the criticism of Tournefort, refused to allow the "Discours" to be published in France. The lecture might have gone the way of Camerer's letter, but Vaillant had friends outside France, the English botanist William Sherard and Professor Hermann Boerhaave at the University of Leiden, and they saw to its publication in both French and Latin. Not long after, Vaillant would write to Boerhaave, with relish, that "our lecture has created a real fracas here."

The lecture became famous. In 1725, when Carl Linnaeus was a student in Sweden, he absorbed its gist and spirit. "The petals of the flower," he wrote, obviously influenced by Vaillant, "in itself contributes nothing to generation, but only serves as the bridal bed, which the Great Creator arranged so beautifully, and garnished with such precious bed-curtains, and perfumed with so many delicious scents, in order that the bridegroom with his bride may therein celebrate their nuptials with so much greater solemnity. When the bed has been so prepared, it is time for the bridegroom to embrace his darling bride, and loose himself in her." Linnaeus would become famous for establishing the convention of binomial Latin names for plants and animals, as well as the "sexual system" for classifying plants. His system, first presented in *Systema Naturae* in 1735, was based on the details of a flower's sexual anatomy, including the number of stamens and carpels and their relative positions inside the petals. It was an admittedly artificial classification method, not a natural one based on the totality of traits shared among species, but it was simple and easy. It also solidified acceptance of plant sexuality.

Vaillant was lucky to have already been elected to the Academy before his lecture. Certainly, he would never have been approved afterward. When he died in 1722, the Academy declined to issue the traditional "éloge" for its departed member. If only Vaillant had lived another twenty years, he would have had the last laugh: Linnaeus's classification system, based on plants' sexual organs, completely eclipsed Tournefort's.

twenty-two

Who Needs Romeo?

By the mid-1700s, most botanists agreed that flowers had male and female organs and engaged in a sexual exchange, but the mechanics of the process were a complete mystery. How exactly did pollen "fecundate" a flower's ova and create fertile seeds?

There was no analogizing from the anatomy and behavior of *viviparous* (live-bearing) animals: No one had yet seen any sort of eggs or seeds in mammals' ovaries. (Karl Ernst von Baer would spot the first mammalian ovum in a dog in 1825 and the first human ovum in 1827.) Chickens and ducks and other *oviparous* (egg-laying) animals seemed to offer the most germane models of reproduction. Semen from roosters and ganders was obviously involved in engendering viable offspring. Exclude the males from the henhouse and you got eggs to eat; let them dally with the females and you got chicks. But how could semen have a fecundating effect when eggs have an impermeable shell?

When it came to plants, there was an additional mystery. Birds have a *cloaca,* a cavity that leads to the urinary and fecal tracts, and, in the case of females, also serves as a pathway for semen to the uterus. But if pollen were the equivalent of semen, then how, after wind dropped it on the stigma of a pistil, did it get to the ovary to impregnate an ovum? The style looked as if it should be a cloacal passageway, but on examination under a microscope it proved to be solid. Pollen grains can't fall down the style.

If semen couldn't penetrate an eggshell and pollen couldn't pass down a style, then, people reasoned, some nonphysical force of semen and pollen must do the impregnating. According to the esteemed William Harvey in 1662, "it is certain that the semen of the male . . . carries with it a fecundating power by a kind of contagious property, [working] in the same way as iron touched by the magnet is endowed with powers and can attract iron to itself." Other observers wrote of a "volatile spirit," a "germinative spark," or an "*aura seminalis*" that operated on the egg.

No one had trouble believing that the contribution of semen or pollen was of such an ineffable nature. In fact, it was hard to see why a male was needed for reproduction at all. Most people believed that inside a bird's egg or inside a plant's seed was a preformed, so-tiny-as-to-be-invisible version of the coming creature or, alternatively, its infinitesimal but as yet unassembled parts. Quite likely, it was thought, semen just jump-started its growth.

There was a second part to what became known as the

"ovist" theory: Inside the ovary of the infinitesimal hen-to-be are eggs that hold all the future descendants of that unborn hen. In other words, ovists believed that the egg of an animal or the seed of a plant is like a set of Russian nesting dolls. Each new generation is inside the other, ready to emerge into the light at its appointed time. As improbable as the theory sounds to twenty-first-century ears, ovism was actually the most reason-based explanation of reproduction. The only alternative theory was *epigenesis,* which held that a complete creature came into existence in an egg or a seed out of only the unorganized, raw materials inside. Epigenesis required a supernatural force to operate every day in every act of conception, magically creating a thing out of nothing. To seventeenth-century rationalists this idea was anathema, reeking of medieval occultism. Instead, they held that God intervened in the world only once, when He created the world in six days. When he created living creatures on days five and six, he simultaneously created all their future descendants, lodging them within the wombs, eggs, and seeds of the first generation. Then, having stocked the shelves, He bowed out and left history to unfold.

In 1677, Antonie von Leeuwenhoek added a new explanation of how conception occurs. Leeuwenhoek was a Dutch cloth merchant who made remarkable, simple microscopes. He fashioned his lenses by heating a glass rod in a flame and then pulling the ends apart to draw out a slender glass thread. After breaking the thread, he touched its delicate tip to the flame so it melted and formed an eighth-inch bead, which he painstakingly polished. Finally, he placed the bead in a hole in a small brass plate. While its field of vision was very

narrow, the magnification was greater than Hooke's microscope. (I tried a Leeuwenhoek instrument at Notre Dame. It
was surprisingly powerful but uncomfortable to use: I had to
hold the plate right up to my eyeball and couldn't blink while
looking through it because my lashes would brush the plate.)

Leeuwenhoek was uniquely successful in making this kind
of microscope. He was also an unschooled man, spoke and
wrote only Dutch, and was paranoid that other, more worldly
men like those at the Royal Society would steal his technology. He was happy to share his drawings of his findings with
those worthies when they asked, but he never responded to
their requests to see his instruments. In 1676, he had sent the
first drawings, which were also the first sightings, of single-
celled microorganisms, which he called animalcules. The following year, taking up one of his best devices, he peered into
a drop of his semen, and a new possibility—or rather millions
of new possibilities—for the source of new life wriggled into
view: spermatozoa. Leeuwenhoek asserted that inside each of
the rounded heads of these new animalcules was a tiny new
human waiting to be born. "It is exclusively the male semen
that forms the foetus and . . . all that the women may contribute only serves to receive the semen and feed it." The ovists'
idea of preformed beings was correct, he and other "sperm-
ists" opined, but they had the wrong idea about their location.
God had put all future human beings not in Eve's ovaries, but
in Adam's testicles.

The spermist theory quickly won adherents among the
natural philosophers. Wasn't it more likely that the more
powerful sex—and the one of which they happened to be
members—is the one that creates new life while females

simply provide a nest? In 1703, Samuel Morland, an English baronet and a remarkable polymath, applied the spermist theory to the vegetal world, advising that pollen "is a congeries of seminal plants, one of which must be convey'd into every ovum before it can become prolific."

Over time, however, doubts about spermism arose. According to Leeuwenhoek, a million human spermatozoa could fit in a grain of sand. That meant that if each sperm held a tiny person, God was horrendously wasteful of creatures made in his own image. It made masturbation mass murder. Pollen, which blew about in spring in quantities great enough to fur a pond in a coat of yellow, was an even larger, if less heart-wrenching, waste of life. While Nature was obviously prodigal of youth—in early eighteenth-century London, almost half the children died before their second birthday— this level of carnage was hard to accept.

Ovism won a new lease on life in 1740 when Charles Bonnet discovered that aphids can reproduce parthenogenetically, that is, without the participation of a male. (The female effectively clones herself.) After the mid-1700s, many scientists concluded that spermatozoa are actually parasites, although why they are found only in postpubescent males was perplexing. Another possibility began to be discussed: Perhaps the fecundating power of semen lay in the liquid in which the parasites swim. The favored theory was that new creatures reside, preformed, in female ova, which are pushed into life by the liquid component of semen.

About this time, the Italian scientist Lazzaro Spallanzani supplied some actual evidence to this heretofore completely abstract debate. Spallanzani was a genial, round-faced, bald-headed man who looked a bit like actor Wally Shawn. Born in northern Italy in 1729 to a lawyer and his well-connected wife, at the age of twenty he embarked on the study of law at the University of Bologna, where his cousin Laura Bassi was the first female professor of physics and mathematics in Europe. Under the influence of the remarkable Bassi, he reoriented his studies to physics, chemistry, and natural history, and earned a doctorate in philosophy. He was also ordained and became associated with two congregations, although the Abbe Spallanzani would never spend much time on priestly duties. Instead, he taught logic, metaphysics, and Greek at the new University of Reggio Emilia, not far from his family's home.

He also read the work of the renowned French naturalist Georges-Louis Leclerc, Comte de Buffon, and his sometime collaborator, the English Catholic priest and amateur biologist John Turberville Needham. These cross-channel collaborators researched and wrote on the subject of spontaneous generation, the theory that animals can emerge from inanimate materials. One hundred years earlier, Francesco Redi had demonstrated that the maggots that appear in putrid meat do not leap unparented into existence, but grow from eggs laid by flies. To Buffon and Needham, however, the fact that complex animals did not appear from nowhere did not exclude the possibility that microscopic life generated spontaneously. They thought it quite likely that animalcules invisible to the naked eye might pop into being out of air. In 1750, Needham reported experimental evidence that he was certain confirmed the immaculate birth of microbes. He had boiled meat broth for ten minutes—boiling was by then a well-known method of killing microorganisms—then poured the liquid into vials and corked them. A few days later, he found microorganisms bustling about in the liquid.

A skeptical Spallanzani repeated Needham's work in the early 1760s, with some crucial adjustments to the Englishman's experimental methods. He put the broth in a flask and boiled it for an hour (after drawing off most of the air inside the flask so the air wouldn't expand during boiling and explode the flask). He also hermetically sealed the flask rather than simply corking it, and set up a control group that he boiled at length but only corked. The broth in his sealed flask remained sterile indefinitely while the corked broth grew cloudy with microbes. Needham was unmoved by the evidence. Such long boiling, he wrote, had killed the broth's "Vegetative Force" and damaged the "elasticity" of the interior air, preventing any new organisms from springing into being. After a number of creative modifications to his original experiment, Spallanzani proved to his own satisfaction, although not Needham's, that microbes do not generate spontaneously from "putridity." Microscopic life, like all life, can only come from life.*

Spallanzani's experiments inspired him to look into other questions related to generation, growth, and reproduction. He studied the ability of salamanders and frogs to regrow legs, and snails to grow replacement heads. (Really, they can.) He managed to isolate a single microorganism in a drop of water and watched how it either budded or fissioned to replicate. Some microorganisms, dried and seemingly dead

---

* Not until 1859 did Louis Pasteur finally scotch spontaneous generation completely.

for years, he found he could bring back to life. He studied spermatozoa of various species intensively, subjecting them to motion, chemicals, fumes, and changes in temperature in an attempt to understand just what these "spermatic worms" were. His laboratory methods were meticulous: He repeated many experiments dozens and even hundreds of times, tried to falsify his hypotheses, tested alternative explanations for results, and used control groups. A persuasive writer, he became one of the leading scientists of the day, and in 1769 he accepted a chair at the prestigious University of Pavia.

His work on sperm led him to the central question of whether semen is essential to fecundation or not. If it is, what exactly does it do? Spallanzani turned to aquatic green frogs for his experimental subjects. No one had properly observed amphibian sex, and everyone assumed, because male frogs clasped females, that fertilization took place internally as it does in mammals. Spallanzani was the first to recognize that this is not so, that the male frog's milt, ejected into the water during frog *amours* while the female emits strings of eggs, is the equivalent of semen. But what exactly did the milt do and how did it do it?

Spallanzani focused on the proposition than an *aura seminalis* fertilizes eggs. Could the *aura* work through the air? He tied strings of frog eggs above a glass dish of frog semen, but the eggs did not turn into tadpoles. Could an *aura* radiate through water? In one of history's most charming experiments, he outfitted dozens of his male frogs in tight-fitting, waxed, taffeta trousers (I imagine pink) and set them swimming with female frogs ready to release eggs. Uninhibited by their formal attire, the male frogs responded as male frogs ought, which Spallanzani knew by examining their trousers afterward. None of the nearby eggs became tadpoles. His data should have led him to conclude that direct contact between sperm and eggs was essential to conception. Instead, he concluded that tadpoles *were already complete* in the eggs and the role of semen was to inspire them to grow. How could this be?

For starters, he was a committed ovist. After having spent years discrediting the myth of spontaneous generation, he found it impossible to believe that a "shapeless body, whether liquid or solid" could become an organized being. There had to be a creature in either the egg or the sperm, and his microscopic inspection of sperm revealed no tiny creatures.

A quirk of amphibian reproduction further led him astray. It happens that when a frog or other amphibian egg is pricked and its surface breached, the egg can respond as if a sperm has passed through its membrane. It begins to divide and soon a tadpole, and eventually a frog—a clone of its mother—develops. Spallanzani, in describing his lab procedures, noted that he used a needle or a pencil to manipulate frog eggs. From time to time, he must have pierced an egg because he described the development of virgin eggs into tadpoles. Although these accidents didn't happen all the time, he couldn't help but be impressed when they did. Frog eggs, he wrote, "are not eggs, as Naturalists suppose, but real tadpoles. . . . The egg is nothing but the tadpole wrapped up and concentrated." Moreover, "There is no essential difference between impregnated and un-impregnated eggs." He then verified his discovery with similar experiments on several species of toad and newts. "The aspersion [sprinkling] of the seed of the male," he concluded, "is a condition necessary to the animation and evolution of the fetus," but not to its original formation.

If frog eggs were really slumbering frog embryos, were plant ova likewise quiescent plant embryos? He ran experiments on hemp, pumpkin, spinach, dog's mercury, and a dozen other species. Like Camerer, he strove to destroy any source of pollen and to prevent any accidental fecundation. Many times he was successful in isolating his subjects, and no viable seeds resulted. But sometimes his female flowers—even if cloistered indoors—*did* produce fertile seeds. Again, quirks of nature were responsible. The tendency of some species, like hemp and spinach, to produce the occasional hermaphrodite individuals resulted in seed set. He didn't know that the female flowers of his pumpkins could be pollinated by other varieties of *Cucurbita,* such as zucchini. In some plant species, distantly related pollen does not fertilize an ovum but can stimulate it to divide parthenogenetically, as the needle provoked the frog eggs. And he had no idea how far pollen of his experimental subjects could travel: hundreds of yards easily and much greater distances if transported on hair or clothes. Spallanzani conducted many of his experiments outdoors, so when he confidently reported that there were no hemp yards nearby to threaten his virgin hemp, who knows what wild and subtle Romeo had breached his convent's walls? In any case, he became convinced—and thanks to his reputation, many others did, too—that pollen, like semen, played only a minor role in creating offspring.

If only Spallanzani had met Josef Gottlieb Kölreuter. Kölreuter was born in 1733 in a small town in the Black Forest region in the southwest of modern Germany, the son of an apothecary. At fifteen, he matriculated at the nearby University of Tübingen, where he studied medicine and botany and published a review of all the experiments to that date on plant sexuality. Degree in hand, he went to St. Petersburg, Russia, in 1759 to work as a natural historian at the Imperial Academy of Sciences. That year, the Academy offered a prize for the best essay providing new evidence on the question of whether plants reproduce sexually or not. Linnaeus, already an ardent supporter of plant sexuality, submitted an essay that asserted that hybrids proved that plants are sexual. (A hybrid is the result of the mating of a male and female of two distinct species.) As proof, he offered up dozens of plants he said had the leaves and epidermis of the male parent and the fruit and bark of the female parent. His essay won him fifty gold ducats.

The problem, or rather the greatest problem, with Linnaeus's essay was that his hybrids were not hybrids. To Linnaeus, any plant that looked like an intermediate between two species was a hybrid, and with those criteria he found hybrids everywhere. (Had he been looking for mammalian hybrids, he might have concluded that the thirty-pound spotted ocelot is a hybrid between a hundred-pound leopard and a domestic cat.) Kölreuter was appalled at Linnaeus's supposed evidence, what he called those "premature births of an over-excited imagination," and launched the first scientific study of hybridization.

He started with tobacco plants, which are hermaphrodites, using the pollen of one species to pollinate the castrated flowers of another. Not all crosses produced progeny, but the successful ones consistently produced offspring that exhibited physical characteristics of both parents. Over the course of six years, Kölreuter made more than five hundred different crosses among 138 different species, repeating the crosses thousands of times, using control groups, and minutely describing the characteristics of the offspring. Between 1761 and 1766, he published four reports on his experiments, reports that should have made it clear that the male parent contributes substantially and in a consistent way to the appearance of its offspring. But Kölreuter's brilliant work went either unread or unappreciated, as Camerer's had seventy years earlier and as Gregor Mendel's would be a hundred years later.

Instead, the old debate about the mechanism of reproduction burbled along. In the first decades of the 1800s, German scientists were just coming into their own, and developed a new theory. Grounded in the new chemistry of Lavoisier, their model for reproduction was a chemical reaction. In plants, according to the German physician and botanist Karl Friedrich von Gärtner, "the liquid in the pollen reaches the ovules after being combined with the liquid secreted on the stigma, so as to give birth there to the embryo." Reproduction in plants was vegetal chemistry.

While the true nature of fertilization remained as obscure as ever, the improved microscopes did uncover an extraordinary secret about pollen. In 1822, Giovanni Battista Amici, a leading Italian microscopist, was closely observing the sticky stigma of a *Portulaca* flower onto which some grains of pollen had fallen and adhered. "Suddenly," he wrote, a grain "exploded and sent out a type of transparent gut" down into the stigma. What he saw for the first time was the pollen tube, which is a cell in a pollen grain that after landing on a compatible stigma, elongates, and burrows its way down the style into the ovary and then into an ovule inside via a pore on the ovule's surface. (Amici couldn't see this, but the tube, which is filled with liquid, becomes the conduit for two sperm cells to swim down into the ovule. One fuses with the ovum to create the embryo; the other unites with two "polar nuclei" to become the endosperm, the tissue that nourishes the embryo.) Here at last was the answer to the impenetrable mystery of how pollen grains pass through the all-too-solid style to fecundate seeds: They tunnel their way in.

Ironically, Amici's discovery led to a revival, or rather a reinvention, of the old spermism. Matthias Schleiden was a young professor of botany at the University of Jena who, in 1838, cofounded modern cell theory, which states that all living things are made of cells and new cells are created by the division of old cells. Marrying his idea with Amici's discovery of pollen tubes, Schleiden suggested that the pollen tube delivers a single-celled embryo to the ovary, where it then divides and grows. "The anther," wrote Heinrich Wydler, professor of botany at the University of Berne in 1839, "far from being the male organ, is on the contrary the female organ: It is the ovary. The grain of pollen is the germ of the new plant, the pollen tube becomes the embryo."

What is most strange about both the ovist and spermist theories is that they flew in the face of what everyone commonly observed: Offspring tend to have features of both their parents. Ordinary folk knew that by whatever mechanism, little Johnny's chiseled looks came from his mother and those long legs came from his dad. Farmers selectively bred animals to produce thicker-coated plow horses, better milk cows, fatter pigs, and more prolific hens. Pigeon fanciers crossbred their birds to create a vast array of sizes, plumage forms, and colors. In 1865, Mendel pointed out (to a completely indifferent world) that if the egg cell of a plant "fulfilled the role of a nurse only, then the result of artificial fertilization could be no other than that the developed hybrid should exactly resemble the pollen parent." But the academics' theories trumped everyday wisdom. "It is the opinion of most physiologists," Charles Darwin wrote in *On the Origin of Species* in 1859, "that there is no difference between a bud and an ovule." In other words, a new plant is in the egg. Pollen simply—somehow—awakens it. In the mid-nineteenth century, nearly two hundred years after Nehemiah Grew first floated the idea of plant sex, botanists were not much closer to an understanding of how it works.

twenty-three

Black Petunias

The only flowering plants my mother ever grew were petunias, which she put in large pots by the front door. They delighted me as a child, with their simple, open trumpets in uncomplicated colors: a pure white, a pink the color of Hostess Sno Balls, a crayon red, and a bright purple the shade of my favorite party dress. It was my job to pinch off the spent blooms, which my mother said encouraged more flowering. The leaves were unpleasantly sticky, but the petals were even softer than velvet, and infinitely lighter in weight. They seemed friendly flowers and eager to please: All summer they blithely spilled flower after flower as if inspired by their own naïve delight in the season.

The petunias I am looking at today in the display gardens of the Ball Horticultural Company headquarters outside West Chicago are not my mother's petunias. No such petals ever unfolded in her pots by the front door. These petals are black, completely and utterly black, without the slightest pentimento of deep purple that you will find in other purportedly black flowers. The only color here, in the deepest depths of its trumpet, is a single dot of brilliant yellow. It is as if all the sun-sent photons that ever had the misfortune to approach this black hole of a bloom had been sucked down its dark throat and were distilled in its depths.

It is a day in late August, and I am in these spectacular gardens waiting to meet Dr. Jianping Ren, the breeder of the black petunia. I had found her petunias at the entrance to the garden, commingled in a large pot with pure white and lipstick red varieties. I can't imagine who would want a garden full of black petunias other than a funeral home operator, but in this mixture the effect is striking, in a Madame X kind of way. Black petunias can be sexy.

Dr. Ren finds me, and together we walk back through the garden and across the lobby. She has agreed to show me her breeding greenhouses, and we'll have to drive about twenty minutes to get there. Ping, as she insists I call her, has thick, black hair cut in a pixie style, a brilliant and ready smile, an energetic stride, and an English that is fluent if just a little eccentric. She is charmingly candid. She is also one of the country's foremost plant breeders and is responsible for Ball's petunia program, one of the company's biggest-selling product groups.

Ping came to the United States after finishing her undergraduate work in 1998 in China. After earning a Ph.D. from Cornell, she started at Ball in 2001, breeding the traditional petunias I knew from my childhood. She joined the long line of breeders who have been manipulating and improving petunias for almost two hundred years. In 1834, a British nurseryman named Atkins created the garden petunia (*Petunia hybrida*) when he crossed two species that had recently been shipped from South America. One, *Petunia axillaris,* has an upright habit and white flowers that have a long, narrow tube, something like the trumpets that medieval heralds played. *Axillaris* emits a scent that attracts its primary pollinator, the hawk moth, which has a long proboscis capable of reaching the nectar produced deep in the tube. The other species is *Petunia integrifolia,* which has a ground-hugging habit, violet-purple flowers with short, wide tubes, and no scent. Its pollinators are primarily bees, which are drawn by color and can maneuver their big bodies comfortably into the wide tube. (The tube also serves as a pied-à-terre for trysting insects.) In the wild, the two species do not produce hybrids even when they grow in the same area. Hawk moths are active only at dusk and, relying on fragrance, rarely stumble into the scentless *integrifolia*. Most bees cannot fit into the narrow *axillaris* tube.

Small bees do occasionally visit both species' flowers during the day, so one would expect that at least a few hybrid seeds would set. But without a breeder's intervention, they rarely do. Why not? The two species are compatible enough that pollen grains of both species send out pollen tubes no matter which of the two species' stigmas they land on. But when both *integrifolia* pollen and *axillaris* pollen fall on *axillaris* stigmas, the pollen tubes of *axillaris* always grow faster and win the race down the style to the ovary. Likewise, on *integrifolia* stigmas, *integrifolia* tubes always beat *axillaris* tubes. This home-field advantage is part of what makes a species a species. There are many other possible barriers to hybridization—incompatible chromosome numbers is the biggest one—but between these two petunia species, the *homospecificity* of pollen is the stopper. Only when a breeder or a lab ensures there is no competition do the two species hybridize.

By 1837, Joseph Harrison was writing in the English *Floricultural Cabinet* magazine that breeders had produced many charming color variations from crosses of the two species made in the greenhouse, including "pale pink with a dark center, sulphur [a deep pink] with a dark center, white with a dark center, and others streaked or veined." Since that time, breeders around the world have tinkered with Atkins's hybrid and made it one of the world's most popular garden plants. By repeated crossings, careful selections, and inbreeding, plus the occasional fortuitous mutation, they have changed the simple purple and white petunias radically. Now some varieties have double the number of petals or edges so deeply frilled that they are hard to recognize as petunia blossoms. Breeders have coaxed yellows out of the gene pool, and fashioned a host of patterns, including stripes, stars, bicolors, streaks, freckles, contrasting edges ("picotees"), and "morns" that slide from hue to hue like a Turner sunset. In 1837 Harrison was impressed by petunia *corollas* (the petals, collectively) that had grown to three inches in diameter; twentieth-century breeders have bred blossoms as broad as seven inches. Petunias now flower more prolifically and branch more abundantly, tolerate drought better, and are more resistant to fungi. Their soft petals, once easily damaged in a hard rain, are sturdier.

Until twenty years ago, however, one trait remained the

same: All petunias on the market had an upright growth habit. Then, in 1995, Ball introduced a new type of petunia, a spreading petunia called Wave. Developed by a breeder at the Kirin Brewing Company in Japan, Wave opened an entirely new market for the petunia as a full-season, flowering ground cover. Put one Wave petunia in a garden and it will cover a four-foot-diameter area in blossoms. In 2001 when Ping arrived at Ball, Wave was where the action was in petunias.

"To be honest," Ping tells me, "when I was assigned to regular petunias, I was a little bit discouraged. Everyone already knows what a regular petunia is. But since it was my assignment, there's got to be something I can do, so I worked really hard."

A new flower color or a unique pattern is just one objective for commercial breeders like Ball. Ball sells seeds to wholesale growers who are keen to buy varieties whose buds open simultaneously. Growers sell to retail garden centers and "big-box" stores that like to have all the colors of a series (say, the heat-tolerant "Madness" petunias) in a blooming display all at once. A petunia that blooms especially early in spring also appeals to growers because they might squeeze in an extra crop that season. They also prize varieties that bloom profusely but remain compact, so they can maximize the number of individuals in expensive greenhouse space. Such factors are equally, if not more, important than new colors to the wholesale growers who buy Ball seed, and Ping found room for improvement.

Ping takes me into one of her light-filled petunia greenhouses where she evaluates new parent lines and promising new crosses. In a commercial greenhouse, the wide waist-high tables (called "benches") are jammed pot to pot, making it difficult to distinguish one individual from another. Here, larger, vibrantly healthy, blooming specimens are generously spaced on long, narrow benches that run the length of the greenhouse. Each specimen has room to demonstrate its tendency to branch and its willingness to flower. The benches are arranged according to color groups and to hues within each group. If you could hover at the ridgeline of the greenhouse and look down, the benches would look like those paint store sampler strips, laid end to end.

We start along a bench devoted to variations on the theme of burgundy, then walk along a bench filled with various shades of magentas. I trail Ping as she walks up one aisle and down the next, along benches of lavenders, royal purples, purplish blues, sky blues, peaches, pale pinks, and hot pinks. She is explaining some of the intricacies of the genetics behind plant breeding—in each pot is a stick with a white card that displays strings of numbers and letters that indicate the plant's heritage—but I find to my dismay that I'm having a hard time listening. The visual stimuli are overwhelming. Actually, it is all I can do to resist stroking the gorgeous petals or, even more tempting, brushing my cheek against them. The amber flowers of one plant are exactly the color of toffee, and I stifle an impulse to nibble one. One piece of information that does penetrate my stupefied brain is that out of every thousand plants Ping produces, 999 will end up in the compost heap. Of these beauties before me, none will be seen by anyone but Ping, her assistants, and now me.

After we pass along the creams and whites—a welcome interlude, like a palate-cleansing sorbet between courses—we stop at a cluster of petunias whose flowers are green. I realize I've never seen a green petunia before. That is not surprising, Ping tells me, because until a few years ago, no one else had, either.

"We must stop here," she says, "because talking about black has to be talking about green" when it comes to petunias. And talking about green petunias takes her back to the spring of 2003, when one of Ball's customers, a wholesale grower in Minnesota, called her up. He had purchased seeds of Ball's white Supercascade, an upright variety with blossoms up to five inches across, but something odd had happened: One of the plants was bearing green flowers. Was anyone at Ball interested in this green petunia?

Ping was. She asked the grower to send her the plant, and found its flowers were a pale lime, almost a yellow green. The color wasn't particularly attractive and the plant's architecture wasn't very good. "It was very loose, didn't branch well," Ping said. Nonetheless, she wanted to acquire this individual's genes, and Ball licensed the plant from the grower.

A green pigment in a plant is obviously not unusual; leaves and stems are green with chlorophyll. The purpose of flowers, however, is not to produce energy but to attract pollinators, and the corolla needs to stand out from the foliage background like a glowing neon EAT HERE sign. Flowers dressed in green petals are generally not dressed for evolutionary success. They are less likely to catch the eye of a pollinator, and therefore less likely to produce offspring. (Wind-pollinated plants, like grasses and many trees, needn't invest in gay apparel.) There is, though, a period when a plant with chlorophyll in its petals might have a slight edge in the competition for survival. Look at a white petunia, Ping said, just as the bud begins to open, and you will see that its immature petals have a green tinge. The immature petals have chloroplasts that are contributing their small bit to the plant's overall photosynthetic capacity. As the bud matures and fully opens, the chlorophyll degrades, and the corolla appears white. In the Minnesotan green petunia, Ping explains, a genetic mutation stopped the chlorophyll from degrading.

Ping started a careful program of crossing the green into various parent lines based on her knowledge of their genetics. At first, nothing interesting emerged, but eventually "all kinds of new colors come out, an incredible color range. Most colors look dirty and are not too pretty—one looked like old blue jeans. But the thing is, many of these colors have never been seen before in petunia."

Petunias inherit two copies of each gene, one from its mother and one from its father. So, which form of the gene (that is, which *allele)* will be expressed in the individual, the mother's or the father's? When Gregor Mendel crossbred pea plants in 1865 and 1866, he focused on two traits: plant height and pea texture. Fortunately for him, the pea plant's height (either tall or short) is attributable to one gene while pea texture (either wrinkled or smooth) is attributable to another. Even better, in both cases one of the two alleles—tall in the case of height; smooth in the case of texture—is dominant while the other is recessive. Whenever a pea plant inherits both a "tall" allele and a "short" allele, the plant's *phenotype* (meaning its appearance) will be tall. A "smooth" allele trumps a "wrinkled" allele. Only when a pea plant inherits two recessive alleles will its phenotype reflect the recessive trait.

Mendel was lucky in his choice of pea plants as experimental subjects. Height and pea texture follow the simplest, "Mendelian" inheritance pattern. It's a good thing he didn't run his experiments by crossing *Petunia axillaris* and *Petunia integrifolia*. Instead of a mix of white- and purple-flowered offspring in a 3:1 ratio, he would have gotten all pink-flowered plants, and might never have formulated his laws governing inheritance. Those nineteenth-century pink petunias arose thanks to "partial dominance," where neither color is fully dominant over the other, so a mixing of characteristics results. On the other hand, if Mendel had crossed Ball's green petunias and certain pink petunias, he might have produced something like the Sophistica Lime Bicolor, a Ball petunia whose petals are piebald, with patches of green and pink. He would have discovered "co-dominance" where both alleles are dominant and both traits are expressed. Stripe, star, and picotee patterns also involve co-dominance.

Breeding for flower color in petunias requires an instinct, an education in floral genetics, and long experience. Flower pigments fall into three major groups: chlorophyll (which produces green), carotenoids (yellows and oranges), and flavonoids (primarily reds, purples, and blues). Among the many flavonoids are also "co-pigments," which appear colorless to us but shift the appearance of other colors. A petunia has about two dozen "structural" genes that direct the multiple steps necessary to produce pigments or co-pigments. It also has another two dozen "regulatory" genes that determine when and how those structural genes are turned on or "expressed." Given all these genes, and the two different alleles present for each of those genes, Mendelian inheritance alone would produce a huge number of corolla hues and shades. Add the impact of partial dominance and co-dominance, and the color possibilities become astronomical. Still more factors influence color, including the shape of the epidermal cells of the petal and the pH of its pigmented cells. Change the pH of a petunia pigment cell by 10 percent and you can change its color from red to blue.

That's just pigment inheritance. We haven't gotten to pattern yet. Consider the effect of a regulatory gene called MIC, which produces co-dominance. MIC turns on the blue-red pigments called anthocyanins. If MIC never gets turned on, no matter what other colors lie in the plant's DNA, its flowers will be white. If, as the petals begin to develop in the bud, MIC activates early in development, you get thin white pinstripes on a dark background. If MIC turns on late, you get a broad, white star pattern. Because MIC is influenced by environmental conditions—heat, light, humidity, and temperature—and because buds on any one plant develop at different times, you may find solid, pinstriped, and star-patterned petunias on one plant.

So, while I may see a simple pink petunia on one of Ping's benches, a complex genetic heritage has influenced its appearance. Make a cross, and a color hidden in a recessive allele and unexpressed in either parent may suddenly appear. Which is just what happened one day in 2005. Ping had placed pollen from a purple-with-bright-yellow-star petunia on the stigmas of burgundy-with-white-star petunia. Of the two hundred seeds that she sowed, one seedling grew up and opened to reveal truly black petals. "When we see it," Ping said, "it is so surprising. It is black, black, black. Nobody can believe it."

When she looked at the parents' pedigrees, Ping realized what had happened. In addition to the genes for burgundy, purple, white, and yellow, the parents' lines contained genes—including the mutation for green—for the entire color spectrum. Through genetic reassortment, this one individual expressed all the colors of the rainbow. With every visible wavelength absorbed, no light is reflected, and human eyes see no color at all.

It took five years to bring Black Velvet to market. Neither crossing the parents of Black Velvet again nor crossing two Black Velvet offspring results in seeds that "breed true," that is, consistently produce seedlings with perfectly black flowers. This means that, so far, the variety must be produced from cuttings rather than seeds, a labor-intensive and more costly method of producing a crop.

At the end of my visit, I tell Ping I am going to buy some Black Velvet when I get home, but she tells me it too late in the season to buy petunias at all. Instead, she gives me a small plant that was headed for composting, and I take it with me in a cardboard box on the plane. During the flight, I see that one of the blooms, soft and deeply black, has fallen off, and I am inspired to dissect it. (To think how far I've traveled that a midnight blossom now stirs in me not poetry but a deconstructive curiosity.) I put on my reading glasses and pull the flower apart. And here is a discovery, at least for me. That brilliant dot of yellow, those photons trapped and blazing in the depths of the flower's black hole? In fact, it is a tight cluster of five, pinhead-size anthers loaded with bright yellow pollen. Nothing magical or mysterious, just petunia genitalia.

twenty-four

The Abominable Mystery

Why flowers, anyhow? Plants began to conquer the land more than 400 million years ago and ruled over it for more than 250 million years without producing a single blossom. Why should they have? Flowers are expensive. Sepals, petals, pigments for color, organic compounds for scent: Creating those fancy clothes and complex perfumes takes a lot of stored energy. Instead of manufacturing flowers, a plant could have used those carbohydrates to make more seeds or grow taller, both proven strategies in the competition for survival. Besides, there seems to be nothing in a gymnosperm that corresponds to flowers. Flowers seem to have arisen out of nothing, sui generis. Nonetheless, blossoms—from the oak's minuscule brown nubs to the green spikelets of rice to the multi-petaled splendor of the rose—appear on at least 75 percent of all the world's plant species.

The why and how of angiosperms, Darwin wrote in 1879, is "an abominable mystery." It is a mystery that still has not been fully solved. Part of the difficulty is that flowers have always been fragile and when they die, they fall apart into easily scattered and perishable pieces. The fossil record of early flowers is therefore exceedingly scant. In recent years, however, evolutionary botanists have come to think the living *Amborella trichopoda* will help solve the puzzle.

You will never find anyone who will sell you a bouquet of amborella. For one, the cream-colored, dime-sized flowers on this knee-high shrub are not much to look at, and will set no lover's heart aflutter. If you should want to buy an amborella plant (say, you want to write about it), you'll also be out of luck. The shrub is rare, and grows naturally only in the cloud forests of the island of New Caledonia in the South Pacific. Only a few American conservatories cultivate amborella, which is notoriously difficult to sustain and flowers unpredictably in captivity. Nonetheless, unprepossessing and persnickety as it is, the plant has attracted a great deal of scrutiny recently. It is likely the closest living relative to the first flowering plant, and according to Harvard professor William Friedman, "a critical missing link between angiosperms and gymnosperms."

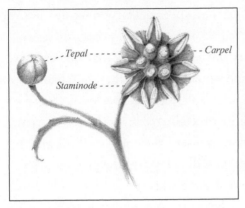

A female amborella flower. The pollen sacs on the female's sterile staminodes (and the male's stamens) are related to the pollen sacs on the scale tips of conifers' cones.

About 140 million years ago, at the beginning of the Cretaceous era, a new type of plant evolved from one of the seed-bearing, nonflowering gymnosperms—mostly conifers—that covered much of the landscape. This first angiosperm was the founder of a new line, the Amborellales. Sometime later, a sister angiosperm line emerged, the Nymphaeales, which evolved to become the modern water lilies. A third line emerged. These, the Austrobaileyales, evolved rapidly and diversified profusely to become almost all of today's 250,000 to 400,000 (depending on who is counting) flowering species, from cucumbers to pansies to elms. The Amborellales, on the other hand, are today represented by a single species, *Amborella trichopoda,* the modern flowering plant least changed from its gymnosperm ancestor. It is a living relic from the age of the dinosaurs.

So, what does *Amborella* have to say about why angiosperms evolved and conquered? As Ping might say, talking about angiosperms is talking about sex. Most gymnosperms are monoecious, although gingko trees, which are one of the few nonconiferous gymnosperms, bear either male or female organs. Among conifers, male and female cones can be on separate individual trees but are more often segregated on either the lower or upper branches of a single tree. (The male cones are smaller and drop soon after they release their pollen. Female cones grow larger after pollination, taking anywhere from a few months to two years before falling.) On the other hand, most angiosperms are hermaphrodites, with stamens and carpels inside the same organ.

Amborella is in a confused state when it comes to sex, as if transitioning from gymnosperm to angiosperm anatomy. It has separate male and female organs on the same plant like many conifers. The male organs have stamens only, but they don't look like modern stamens, that is, with filaments topped with two pollen-bearing sacs. Instead, the two pollen sacs are carried on the edge of flat and broad petals that look very much like the scales in male conifer cones. But the amborella also has organs that *look* like hermaphrodites, with both carpels and stamens. These stamens, however, are sterile, making them *staminodes*.

How did these other organs (which function as females) manage to acquire both carpels and nonfunctional stamens? The debate is far from over, but it seems the first and critical step was a "whole genome duplication" in the common ancestor of all flowering plants. A gymnosperm (like a human) is *diploid,* which means it has one copy of its genome in the nucleus of all of its cells. (The term *diploid* indicates that the genome is a set of paired chromosomes, one from each of its two parents.) After the whole genome duplication, the mutated plant and its *tetraploid* descendants had two copies of its genome in the nucleus of its cells. These plants became instant winners in the survival game. They now had spare genes for experimenting, spare genes that might mutate freely without endangering the viability of the individual. If a mutation in a spare gene conferred an advantage, the mutation could be passed on. If the spare gene mutated in an unproductive way, it would be lost.

Today, most flowering species have more than one copy of its genome, and thousands have a dozen or more copies. Plants with multiple copies are called *polyploids*, and tend to be larger, have more complex flowers, and bear bigger fruits than their genomically less endowed relatives. Humans have shaped plant evolution by selecting polyploid plants for cultivation. Modern garden strawberries, for example, are octoploids and some have berries so large they need to be cut in quarters to be eaten, while their wild relatives are diploid or at most tetraploid and have berries the size of a pea. Plant breeders often propagate garden flowers with doubled genomes to produce double the usual number of petals. In the era just before the amborella emerged, however, a doubled genome simply meant a new plasticity for basic forms and functions. That plasticity was critical to evolving leaves into colorful petals and into *sepals,* the leaves that protect unopened buds. Gene redundancy also was a factor in developing male and female organs in the same flower.

Gender in gymnosperms of the era must have already been nuanced and mutable. Only four divisions of gymnosperms are extant, but some of their members have interesting sexual variations. Some gymnosperms change genders as they mature. Male gingkoes occasionally metamorphose into females, much to the distress of city residents. (Gingkoes are popular urban street trees—they're so tough some survived the nuclear blast zone in Hiroshima—but female seeds are terribly messy and their odor has been compared to vomit or rancid butter.) The Mediterranean cypress starts off female but later becomes male. Many pines have their male cones on the lower branches and female cones on the upper, but several varieties are sexually upside down. So, when the gymnosperm genome first doubled and was chock-full of extra and as yet unassigned genes, gender would have been particularly malleable. Amborella, with its organs that are unisexual in function but hermaphrodite in appearance, is not a shocking experimental result. And amborella is known to go transgender, too.

One of the most important innovations of amborella and later angiosperms was to enclose its seeds in an ovary. No gymnosperm seed has an ovary: Their fertilized ovules are "naked," covered only in a single, thin protective layer. An angiosperm's ovary protects its embryo from desiccation and physical harm. Some species' ovaries enlarge significantly to become an edible fruit. If you're a plant and you want to have your seeds spread widely, fruits are the way to go. Animals transport them inside their digestive tracts, and then deposit the embryo far from the parent, along with a nice pile of nitrogen-rich manure. Botanically speaking, grains and nuts are also fruits. It's just that their ovaries are dry and tough.

So where did the ovary come from? Gene duplication again. The ovary is the bottommost part of the carpel, and the carpel, genetic analysis shows, is a repurposed leaf, which by folding and fusing its edges together came to form a second, impermeable covering over the seed. In fact, the amborella carpel has a visible seam where the leaf has not quite fully fused. An edible fruit is not the only transportation method that angiosperms developed for their seeds. Thanks to those duplicated genomes, external seed structures such as burrs and hooks, plumes, and sticky liquids have evolved to send seeds far down the road. Wait a few million years, and angiosperm seeds will be hailing cabs.

About 80 percent of angiosperms are pollinated by insects; most of the remainder use wind to transport pollen. I assumed wind pollination evolved first since that seems both the simplest method, as well as the way gymnosperms get the job done. Not so. In the Carboniferous era, insects were already crawling into gymnosperms' male cones in search of pollen, which is a highly nutritious food. These ancient thieves didn't waste time and energy visiting pollenless female cones. The innovation of the amborella and later angiosperms was to put the thieves to work as go-betweens, connecting male stamens and female carpels. Although the staminodes in the amborella's female flowers are sterile, they act as decoys, attracting hungry insects who, having visited the real male flowers, inadvertently drop off pollen on the nearby stigmas.

The insects that crawled into the first amborellas 140 million years ago were searching only for pollen and had no expectation of finding nectar. Nectar and nectaries, which are specialized glands found at the base of flowers, had yet to evolve. Darwin might have considered nectaries to be another abominable mystery, since they also seem to have come out of nowhere. But nectar is made of concentrated sugars manufactured in leaves and distributed throughout a plant via phloem. Phloem (and not, to my surprise, xylem) also bring water to flowers. It is thought that phloem at the base of the flower, under the increased pressure of water that accompanies flower development, sometimes leaked a little sap. You might think leaky phloem at the base of flowers would be a disadvantage. After all, what is draining away is hard-earned energy that the plant could use for growth and repairs. But, like the leaky phloem at root tips that succor mycorrhizae, leaky phloem in flowers had an offsetting advantage: Insects favored flowers that provided not only a pollen dinner but a sweet postprandial drink. Over time, as plants whose flowers always kept a well-stocked bar prospered, mutations in nearby structures evolved into nectaries. Ever more attractive petals and scents evolved, too, to ensure that the location of the restaurant would be no mystery, abominable or otherwise.

twenty-five

Cheap Sex

Of Darwin's voyage on the *Beagle* and his publication in 1859 of *On the Origin of Species*, the foundation of evolutionary biology, much has been written. Less well-known is the fact that after he published the *Origin* at the age of fifty, he directed much of his intellectual effort in his remaining years to the study of plants, writing five major books and some seventy articles on botanical subjects. Had he never developed his understanding of evolution, he would still have been renowned as a major figure in the history of botany.

Darwin's botanophilia was born at his childhood home, the Mount, in bucolic Shropshire. The Mount had extensive gardens where he and his five siblings played and an orchard of scientifically bred apple trees. His mother, who died when Charles was eight, nurtured tropical species in a heated greenhouse. In a portrait painted in 1815, a six-year-old, rosy-cheeked Charles—the beetle-browed scientist with the Mosaic beard far in the future—kneels with his arms hugging a big red pot of yellow flowers. Making collections of natural objects was a fad of the period, and Charles became an avid collector of beetles. He also claimed to have had an early curiosity about the way plants work, and would later recount how he told another little boy that he "could produce variously coloured polyanthuses and primroses by watering them with coloured fluids."

Charles started his university career in medicine at the University of Edinburgh, but the gore and mayhem of the operating room horrified him, and he switched to Cambridge to prepare for the ministry. (This goal he later judged ironic, "considering how fiercely I have been attacked by the orthodox.") He was hardly a devoted student during his three years at the university, admitting that he "got on slowly" in math, had forgotten his prep school Latin and Greek, and his attendance at most lectures was "nominal." The only course he attended regularly, and he took it at least twice, was that of John Stevens Henslow, clergyman and professor of botany. "I became well acquainted with Henslow," he wrote, "and during the latter half of my time at Cambridge took long walks with him on most days; so that I was called by some of the dons, 'the man who walks with Henslow.'" Under Henslow's influence, his former passion for beetling cooled and he took up "herborizing" with avidity. He began to imagine his future self as minister of a country parish, a position that appealed primarily because it would leave him plenty of time to pursue botany and his other naturalist interests. First, though, having read and reread the German explorer Alexander von Humboldt's 3,754-page account of his travels through the wilds of South America in the early years of the century, he wanted to go on a voyage of discovery. He schemed to get to the Canary Islands. When that trip fell through, Henslow helped him get hired as a naturalist and companion to the captain of the HMS *Beagle,* who was charged to collect navigational information about the world's coastlines. The ship set sail in 1831.

In September 1835, near the end of the five-year voyage, Darwin arrived in the Galapagos Islands, off Ecuador in the Pacific Ocean. Henslow had taught him that oceanic islands were often home to unique collections of plants, an interesting phenomenon that the reverend attributed to the work of the divine Creator. Indeed, as Darwin traveled around the Galapagos, he saw vivid demonstrations of what we know as "island endemism." He assiduously collected blooming specimens, per Henslow's instructions, carefully labeling them according to where he'd found them. Island endemism, he noticed, didn't appear to be limited to plants; the mockingbirds also looked different from island to island. Although he was unsure whether these were varieties of a single species or separate species, he was already mulling over the significance of their differences.

Darwin returned to England in October 1836 and handed over his collections to specialists for identification. Within months, ornithologist John Gould told him that his Galapagos birds that he had thought were blackbirds, grosbeaks, finches, and wrens were all different species of finch. The mockingbirds were not varieties, but separate species. "In October 1838," Darwin wrote in his autobiography, "that is, fifteen months after I had begun my systematic enquiry, I happened to read for amusement Malthus on Population, and being well prepared to appreciate the struggle for existence which everywhere goes on from long-continued observation of the habits of animals and plants, it at once struck me that under these circumstances favourable variations would tend to be preserved, and unfavourable ones to be destroyed. The result of this would be the formation of new species." Two years later, he was well on his way to an understanding of how natural selection works on variations to create, modify, and extinguish species.

Something, however, troubled him about evolution and plants. The operation of natural selection presupposed a supply of individuals with variations. Those variations, he knew, resulted from the crossing of male and female individuals who looked different from each other and produced offspring that looked different from their parents. But if this was true, the common understanding of the way sex worked in flowering plants must be wrong. Everyone assumed, and Henslow had taught him, that a hermaphroditic flower fertilizes itself. Why else would its carpels and stamens be mere millimeters apart inside their petals? But, Darwin realized, if individual flowers fertilized themselves, then the seed would yield offspring identical to the parent. (He was unaware that random genetic mutations also create variation.) There would be no variation among individuals, and no individual would be any more or less fit than any other. There would therefore be no natural selection, and evolution of species would not occur. Clearly, this had not happened; anyone could see that flowering plants are rampantly speciated. So, Darwin had to reconsider the nature of sex in flowering plants. During the summers of 1838 and 1839, he later wrote, he "was led to study the cross-fertilisation of flowers by the aid of insects. . . . My interest in it was greatly enhanced by having procured and read in November 1841, through the advice of Robert Brown, a copy of C. K. Sprengel's wonderful book, *Das entdeckte Geheimnis der Natur.*"

This inspiring book (whose full title in English is translated as *The Secret of Nature in the Form and Fertilization of Flowers Discovered*) had been published nearly fifty years earlier, in 1793. Its author, Christian Konrad Sprengel, was forty-three years old at the time, living alone in attic rooms in Berlin and surviving on a small pension. A year before, he had been the director of a school. After his doctor advised him to take up outdoor activities for his poor health, he had become fascinated with flowering plants in general and obsessed by the puzzle of pollination.

It was understood at the time, thanks to Kölreuter, that insects helped fertilize flowers, but Kölreuter believed that their visits were haphazard. Insect pollination, he thought, was an accidental occurrence and certainly not essential for the propagation of the species. It might be more important for dioecious species whose male and female flowers were separated, but for hermaphrodites, he assumed it was a secondary strategy. What Sprengel discovered was that there was nothing accidental, nothing casual, about insect pollination. In fact, "nature had arranged the whole structure of the flower for this method of fertilization." For one, he realized that "those flowers whose corolla is differently colored in one place than it is elsewhere always have spots, figures, lines, or dots of particular color where the entrance to the nectary is located." These "nectar guides" are a method of leading the insect to its reward. Fragrance is another guide. Night bloomers, he wrote, have "a large and light-colored corolla so that they catch the eyes of insects in the darkness of the night. If their corolla is inconspicuous, then this short-coming is substituted by a strong scent."

The ins and outs of pollination had not been Kölreuter's primary interest, and his observations had not been rigorous. He had examined only a handful of species, but Sprengel studied nearly five hundred, observing many of them continuously for days in their natural habitats to catch "nature in action." Pollination can occur at dusk or at night or only once for a few seconds in the space of several days. (Orchids can remain in bloom for months because their highly particular pollinators take so long to find them.) Gnats and tiny flies can be stealthy pollinators: You need to practically put your eye to the blossom, or you may miss the event. But watch long enough, Sprengel knew, and a pollinator almost always shows up. Kölreuter had also been misled by plants that developed seeds even without an insect visit. He hadn't thought to test the seeds' fertility, but Sprengel did, and found them sterile. Insects are essential to angiosperm reproduction.*

For all his trouble, Sprengel was rewarded with a revolutionary discovery. In nature, a hermaphroditic flower "cannot be fertilized by its own pollen but only by the pollen

---

* A relatively few plants—peas and peanuts, for example—do self-fertilize. Others, like the soybean, do so as a fallback method after cross-fertilization has failed. (The "nut" of the peanut plant, by the way, is actually a pollinated ovary of a flower whose stalk continues to grow while bending downward until the ovary is pushed into the ground, where the fruit ripens.)

of another flower." Anatomy ruled against self-fertilization, for one. Despite their proximity in a flower, anthers and stigmas do not routinely touch each other. Often they have mismatched heights that reduce the likelihood of contact. Timing is another issue: A flower's anthers ripen and release their pollen, and only then do its stigmas become receptive, or vice versa. This nonsimultaneity, or *dichogamy*, prevents self-fertilization.

Sprengel also put to rest—or would have put it to rest if anyone had read his book—some far-fetched theories about nectar. Some botanical authorities had asserted that the purpose of nectar is to feed the developing seeds in the ovary, and claimed that insects harm the flower by stealing it. Others thought nectar was a danger for flowers, and that bees removed it in order to protect their source of pollen. If nectar wasn't collected, it was thought to accumulate, thicken, and destroy the development of the fruit. Sprengel knew otherwise. "The nectar is for the flowers what a spring is for a clock. If one takes the nectar from the flowers, one renders all their remaining parts useless; one thereby destroys their final purpose, namely, the production of the fruit."

Botanists of Sprengel's generation simply couldn't absorb his unsettling ideas, no matter how much data he provided. This obscure amateur was proposing that the Creator puts male and female organs in the same flower—a perfectly sensible arrangement for beings that can't get up and move to find their mates—and then prevents their union? Absurd. To top it off, He then concocts a labyrinthine system in which insects on the hunt for nectar inadvertently deliver some distant flower's pollen? The theory seemed ridiculous. Sprengel didn't help his cause because he couldn't propose any purpose for this convoluted method of congress. Besides, this business of cross-fertilization reminded people of hybridization, and everyone knew that hybrids, such as donkeys, were often infertile. One expert called his theory of flowers "an amusing fairy tale." His book was never translated into English (and still isn't in full), and he never wrote the second volume he intended. "Poor old Sprengel," as Darwin called him, died in 1816, his reputation so obscure that to this day no one knows where he is buried.

Darwin was perfectly primed to appreciate Sprengel's work; the German amateur confirmed what he already suspected. Sprengel's data, along with the information Darwin collected on his voyage, the tables on genus/species ratios he derived from the scientific literature, conversations with animal breeders, and examples gleaned from his reading went into refining his theories and into the *Origin,* which he started drafting in 1851. Eight years later, when the book was published, he had moved far beyond Sprengel. Sprengel had recognized that "Nature abhors perpetual self-fertilization," but didn't know why. Darwin did. Cross-fertilization results in more vigorous and greater numbers of offspring than selfing. The more vigorous offspring of crosses pass down the inherited traits, including flowers whose carpels and stamens mature at different times, that contribute to their superior survival rates.

Darwin rushed the *Origin* into print, spurred by the knowledge that Alfred Russel Wallace was readying his own manuscript proposing a very similar theory. After publication, he was swamped with correspondence about the book and overwhelmed by the constant need to answer questions about it. In 1860, he turned to a study of plants, and orchids in particular, not only for psychological relief, but also to collect more evidence for his theory. In his book on fertilization in orchids, *The Various Contrivances by Which Orchids Are Fertilised by Insects* (1862), he would write, "It is an almost universal law of nature that the higher organic beings require an occasional cross with another individual. . . . Having been blamed for . . . propounding this doctrine without giving ample facts in [the *Origin*], I wish here to show that I have not spoken without having gone into details."

Wild orchids grew in profusion in the countryside near Darwin's home southeast of London, and he began digging them up to transplant in his garden. After fully investigating British species, he branched into the more elaborate tropical ones. Expensive glass conservatories had become a status symbol among the wealthy in nineteenth-century Britain, and gathering a collection of exotic orchids was an upper-class hobby. Some aristocrats went so far as to hire a collector to scour the tropics for them. The socially well-connected Darwin reached out to those who might send him unusual specimens.

Although he accumulated dozens of species, these were only a tiny fraction of the *Orchidaceae,* one of the largest botanical families. About twenty-five thousand species inhabit every ecological niche, including underground, of every continent except Antarctica. Orchid flowers are astoundingly diverse: They look like jugs, slippers, bees, strange sea anemones, spiders, frizzled ribbons, ducks in profile, long-eared donkey heads, as well as jasmine, crocus, violets, and pea flowers; they come in every color and color combination except true black; and they can be scentless or stink of rotting meat or waft intoxicating perfumes. The plants range in weight from less than an ounce to, in the case of the *Grammatophyllum papuanum,* a ton. Orchids seemed to fly in the face of evolution by natural selection. How could such fanciful structures have anything to do with the tough business of survival? As Thomas Huxley, Darwin's friend and a chief evangelist for his theory, asked, "Who has ever dreamed of finding an utilitarian purpose in the forms and colours of orchids?"

Darwin did, and he made a convincing case that orchids are a striking example of "descent by modification," an iterative process in which a small adaptation leads to a slight improvement in successful fertilizations, increasing the numbers of offspring having that adaptation and spreading the adaptation throughout a population. Despite the variety in orchids' appearances, all are built on a basic anatomical plan: three sepals, three petals (one of which is the *labellum,* or lip, where the pollinator lands), and the column. The column is a single, fingerlike organ that is home to both the stigmatic surface and packets of bright yellow pollen (*pollinia*) that protrude outward from atop slender stalks. The orchid's strategy is to attract an insect to enter a flower in such a way that it is tagged with packets of pollinia. The stalks and their packets adhere to a body part—head, abdomen, back, or proboscis—of the visitor. Within seconds, the stalks wilt or twist in such a way that when the insect enters another flower, the pollinia are precisely positioned to miss the male portion of the column and contact the sticky stigmatic surface, where they detach.

Darwin realized that every seemingly useless ridge and fold of the flowers, all their colors and markings, their scents, and all their odd projections have been shaped by natural selection to serve a reproductive function.

So convinced was he of this arrangement that he confidently offered a prediction of the existence of an exceedingly strange and unknown moth. A friend had sent him several specimens of the star orchid of Madagascar, *Angraecum sesquipedale,* whose six-inch flowers have waxy white petals, a strong, spicy fragrance, and, at their base, a "green, whiplike nectary of astonishing length." What insect, he wondered, could possibly reach the nectar at the bottom of this narrow, curving, twelve-inch-long nectary? He tried maneuvering needles and bristles into the flower, but with no success. Only by poking a wire down the full length of the spur was he able to reach the nectar. Given that the flower is white and has a pungent smell, he knew that the pollinator had to be a moth, and that it would have a tongue a full foot long. He also predicted that the moth would be large; he had found he had to exert a significant pressure to push down on a stigma in order to drop off the pollinia.

His prediction was ridiculed by entomologists who had never seen a moth with a tongue anywhere near that length, but in 1903 just such a moth was discovered. The *Angraecum sesquipedale* orchid is pollinated by a brown hawk moth with a five-inch wingspan and a twelve-inch tongue. The tongue is usually curled up in a tight coil, but when the moth sees or smells its particular flower, liquid streams into the interior of its tongue, unwinding it like a party horn. The moth was named *Xanthophan morganii praedicta.*

The orchid *Angraecum sesquipedale* and its pollinator
Xanthophan morgani praedicta.

Darwin thought orchids always provided a payoff to their pollinators in the form of nectar, and although he had heard of reports of nectarless flowers, he didn't give them credence. Surely, the insects would learn not to waste their time visiting rewardless species. Any insect able to spot the species with empty flowers and avoid them would leave more progeny and spread its lie-detecting ability throughout the species.

In this instance, Darwin was wrong, underestimating the precision of orchid mimicry. About one-third of all orchid species are cheaters, promising lunch but delivering nothing. There are orchids that mimic the appearance of nectar-filled flower species of completely unrelated genera. Donkey orchids, as John Alcock demonstrates in his delightful book, *An Enthusiasm for Orchids: Sex and Deception in Plant Evolution,* have evolved to look astoundingly like members of the pea family. Pink Enamel orchids appear to have five identical shiny, bright pink petals, and look like a number of meadow wildflowers that offer nectar. These orchids may not fool all the bees all the time, but they don't need to. Inside their pollinia are millions of tiny pollen grains, far more than in a pea flower's anthers. Orchids need dupe only a few bees to arrange transportation for their bountiful pollen. Meanwhile, they preserve their energy by forgoing the manufacture of nectar.

The bee orchids (members of the *Ophrys* genus) are equally deceitful, but instead of a false promise of food they dangle the possibility of sex. An *Ophrys* labellum has evolved to resemble the backside of a female bee whose head appears to be plunged into the flower. The male bee, spotting what looks like a bulbous, furry bee bottom and getting whiffs of a specific fragrance that telegraphs "receptive virgin female who is exactly your type," pounces and tries to copulate. (If you want to take a walk on the wild side, you can find clips of bee and *Ophrys* pseudocopulation on YouTube.) After about a minute of unsatisfactory lovemaking to a flower, he gives up and flies off in search of a more satisfying partner, unwittingly carrying away orchid pollen.

Research shows that he flies some distance away before trying again. Why doesn't he just try the next flower on the next branch? The answer is not because the embarrassed bee wants to avoid any witnesses to his previous humiliation. The reason lies in the evolution and complexity of orchid fragrance. Each *Ophrys* species has evolved to produce the precise mix of hydrocarbon compounds that imitates the dozen or more compounds produced by the female of only one pollinator bee species. In this way, the orchid ensures that the frustrated male bee will seek another blossom (disguised as a bee) of its own species, carrying its pollen only to where it can be effective. But if the frustrated male bee were to move only to another flower of the same plant, the orchid would not have accomplished its goal of cross-pollination. So, the orchid has two other tricks up its flower to ensure that the lover moves farther on. Researchers at the Institute of Zoology in Vienna have discovered that after *Ophrys sphegodes* has been pollinated by its pseudocopulating pollinator bee, the flower immediately produces a new fragrance compound that is the faithful duplicate of farnesyl hexanoate, the compound that female bees emit after they have successfully mated. A whiff of that pheromone in the vicinity quickly sends the male bee on his way. In other orchid species, the male bee can sense subtle variations in the mix of the sex-related compounds produced by *individual* plants—perhaps a few more molecules of compound number 8 and several fewer of compound number 12—and he avoids the flowers of the plant that have already proved so disappointing.

Why doesn't natural selection make history of males that waste their time trying to tryst with a flower? It appears that the male bees are caught in a bind. Those who are quickest off the mark and beat out their rivals for real virgin female bees are most successful in passing their genes to the next generation. Since there are far more virgin bees than deceitful orchids, on balance male bees who make the occasional mistake of mating with a flower prevail over those who dillydally and discriminate. The orchid is quite satisfied by the amorous bee. It has attracted an avid pollinator highly likely to carry its pollen to exactly the right flower at minimal cost, just a sexy skirt and the right perfume.

twenty-six

Scent and Sex

Darwin couldn't appreciate the exquisite precision of orchids' mimicry of scent. It wasn't until the mid-1970s that scientists were able, with the use of a gas chromatograph/mass spectrometer, to tease apart the individual organic chemical compounds of fragrance and measure the volumes of those compounds. While a rose by any other name still smells as sweet, horticultural scientists now can define every molecule that creates that sweetness. There is no longer anything ineffable about the fragrance of a rose or an orchid, or, as I learn from Professor David Clark at the University of Florida, a petunia.

I meet up with Dave in August in one of the experimental greenhouses behind the building that houses his office and labs. Dave sounds like Appalachia and looks like a man who spends his time with old-fashioned petunias; he's a cheerful man with a healthy pink complexion. He heads a research group at the university dedicated to improving floral and vegetable crops through a combination of biotechnology and conventional breeding. Petunias are a major focus of his work, and we start by looking at the plants he's working on. Petunias can't take the heat of August in Florida, so inside the greenhouse the air is cooled to about 80 degrees. I'm expecting a glorious display, but all the flowers trembling slightly in the light breeze have white petals and long, narrow throats. They're nice, but hardly the first plant I'd put in my shopping cart at the garden center.

Aesthetically, Dave's petunias may not be an exciting group, but scientifically they are. In the last decade or so, the garden petunia has become a "model species," a species that researchers use to study the biology of a broader group of organisms. The petunia is in the same family (Solanaceae) as tomato, potato, tobacco, eggplant, and pepper, plants that constitute the most economically valuable group of agricultural plants other than row crops like corn, wheat, and soybeans. What scientists learn by delving deeply into petunias can improve crops that nourish the body as well as the soul.

The petunias Dave grows are *Petunia* x *hybrida* cv "Mitchell Diploid," the equivalent of the medical researcher's white mouse. Much of the "Mitchell's" genome has been decoded, making it easy to silence or add genes and explore their function. "Mitchell," like the wild *Petunia axillaris* it closely resembles, has big flowers that produce a lot of volatiles. (Volatiles are substances that easily change from liquid to vapor: all the better to smell you, my dear.) This makes the variety an ideal subject for investigating the biochemistry of fragrance.

The scent of "Mitchell" is made up of a dozen or so complex compounds, including clove oil, a sort of medicinal root beer scent, wintergreen, and rose oil. Rose oil is hands-down the most important volatile in the cosmetics and perfume industries. Sad to say—and no surprise to anyone who has been given a bouquet of hothouse roses or bent over to sniff a garden rosebush—the genes that code for rose oil, while still in the genome, are often no longer expressed. The oil used in expensive eye cream or Chanel No. 5 comes from cabbage roses and damask roses, which are cultivated primarily in Bulgaria and sell for about four thousand dollars per pound. It isn't that commercial rose breeders have purposefully eliminated scent from roses, but in selecting and breeding for color, form, longevity, disease resistance, and a host of other traits, scent has been inadvertently lost.

None of the edible Solanaceae have flowers with appreciable fragrance, but you will find fragrance compounds in their fruit. The same molecules that drift into the air to draw pollinators to petunias reside in eggplants, peppers, and tomatoes. What are fragrance molecules doing there? Fruits evolved to attract animals that disperse their seeds, and sugars and proteins are a significant part of the attraction. Some of those proteins contain essential amino acids, "essential" meaning that animals cannot synthesize them and so must eat them to survive. Phenylalanine (fen-l-AL-uh-neen) is one such amino acid. It is also a component of many compounds in petunia fragrance and in tomatoes. In fruits, color says "I'm ripe" and fragrance says "I'm good for you."

Of course, it is not the scent per se of the tomato that draws you to eat one, but the taste you anticipate. Taste is in large part a function of fragrance because our tongues perceive only the chemical compounds of sweet, sour, bitter, salty, and *umami* or savory flavors. (Pinch your nose closed, and you won't be able to tell the difference between a puree of strawberry and a puree of pineapple.) One of the components of a tasty tomato is rose oil. Dave's lab helped identify the gene for rose oil in an heirloom tomato and managed to engineer it into "Mitchell" petunias. The petunia flowers came up smelling like roses.

This snipping and splicing of genes creates a transgenic or genetically modified (GM) plant. In the mid-1980s and 1990s, it seemed that the future of agriculture would be in GM plants. Scientists had visions of stitching vitamins, amino acids, and all sorts of other goodies into the genomes of fruits and vegetables. The breeding limitations imposed by the genetic differences that make a species a species were going to be history. Plant breeding would take place in petri dishes. "Lots of traditional breeders happened to retire at that time," Dave says, "and were replaced with molecular biologists who were cloning genes and creating plants that jumped natural species barriers."

It didn't work out that way. Some major crops were indeed engineered to resist herbicides, viruses, insects, freezing, and drought, but genetically manipulated food, it turned out, makes some people very uneasy. In the United States, regulations were put in place to help ensure that transgenic crops don't compromise health and are safe for the environment. Whether the regulations achieve these goals and whether GM crops really hold any danger is a matter of much debate, but there's no doubt that the expense of the research and the documentation required to meet regulations at various federal and state agencies is significant. According to the World Bank, the cost of preparing a regulatory application for one transgenic corn variety ranges from $6 million to $15 million. To gain approval of even a nonedible crop, say a pink "Rose Perfume" Wave petunia that has a tomato gene inserted to enhance its scent, Dave tells me, would cost a breeder a million dollars—and that would be the expense for just that particular pink. The breeder would have to get approvals for each of the twenty-plus colors in the Wave series that contained the spliced gene.

"The cost of the R&D and the regulations can pay off if you're talking about millions of acres and billions of dollars," Dave says. "But the market value of the entire petunia crop in the U.S. is only three or four hundred million dollars, and that's at the retail level. Plus, the average market life of a petunia variety is five, six, maybe seven years. Then, consumers want something different. Anyhow, what's the dollar value of a pink petunia with a better scent? You just can't get stakeholders interested." In general, transgenics only make sense for commodity crops. Improving flowers and vegetables is still the province of conventional breeders.

Scientists, however, have changed conventional breeding, now that they are armed with the tools of molecular biology and data from genomic sequencing. Say you are a breeder and you have a patent on a white petunia with a minty fragrance and a striking pink-and-green-star petunia with no fragrance. It occurs to you that if your pink-and-green petunia had a mint scent, it would sell well. Thanks to Dave's lab, you know that the white petunia makes ten fragrance compounds, including minty alcohols, and in what proportions. You also know which of its genes are responsible for making each of those scents and where on the petunia chromosome each of those genes lie. Now, in your lab, you can snip out those genes with enzymes and multiply them with recombinant DNA technology. Then you can insert those genes into bacteria of the genus *Agrobacterium*. Finally, you infect your pink-and-green petunias with the bacteria, which transfer fragrance genes.

Once you are satisfied with your transgenic petunia, you do not plant it. Instead you grind up its leaves, subject them to DNA analysis, and see roughly where those white-petunia fragrance genes have landed in the green-and-pink petunia genome. Then you can start creating the old-fashioned way, that is, by transferring pollen from white petunias to stigmas of pink-and-green-star petunias (and vice versa) and then collecting seeds from pollinated flowers. In the past, you would have had to plant thousands of seeds, then water and fertilize and devote air-conditioned greenhouse space to them for two years until the plants flowered and you could discover which hybrids looked beautiful and smelled minty. Today, however, you need only wait five weeks for the seeds to sprout, grind up a little leaf from each seedling, and conduct DNA analysis to see which individuals have the DNA that matches your lab-created transgenic. You still have to grow those individuals to see which branch most prolifically, have the most blossoms, set the most seeds, etc., but you have reduced the time and cost of development radically.

Clark and his colleagues are taking the process a step further. In 2009, the university established its Plant Innovation Program, which brings a group from Dave's lab together with food scientists, psychologists, and marketing experts to change the front end of the breeding process.

"Instead of breeders saying we're going to make this big, bold, yellow flower because we think consumers are going to buy it, we're trying to figure out what consumers prefer. In the case of tomatoes, we're finding out what flavors they want."

The process has already resulted in a new tomato. The Innovation group retrieved heirloom types and had volunteers sample, rate them, and explain their ratings. Then they figured out the biochemical components of those tastes (that is, their fragrance compounds), identified the genes that code for those chemicals, developed a transgenic model, and then bred it conventionally. One such tomato, called Tasti-Lee, was introduced in 2011. Other tomatoes, some of which will be designed to meet different regional preferences, are on the way. New flowers are coming, too, but flower preferences are more complex because color influences expectations for scent. An orange petunia, for example, with wintergreen notes would be weird, but a rose-scented pink petunia—or a Black Velvet that emanates the smokiness of scotch—might be appealing.

As we leave the greenhouse, Dave offers to show me his coleus breeding operation. I'm not especially interested. For one thing, I have a plane to catch. For another, coleus are cultivated for their leaves, not their flowers, which are inconspicuous, and I'm here to finish my research on flowers. Finally, I do not like coleus. I remember coleus as small plants in melancholy combinations of maroon and dark green. Nonetheless, Dave seems very enthusiastic and since I appreciate a good enthusiasm (having indulged in a number of them myself), I agree.

Dave's coleus grow outdoors in breeding lots nearby. As we walk there, I ask him how he came to horticulture, whether he'd always had an interest in plants.

He laughs and says, "I'm originally from east Tennessee. My parents were the generation getting out of the coal mines. My dad worked in a chemical factory and my mom was a seamstress, but we had a farm on the side. My dad used to drop me off on the side of the road to sell beans. I didn't ever mind working in the garden, but it was always work. The whole idea was to get *off* the farm."

Dave and his older sister were the first in their family to go to college. At the University of Tennessee, he started out majoring in engineering—that was what his sister had done and she'd gone on to a successful career at Hewlett-Packard—but he hated it. One day, he decided college just wasn't for him, signed up at a local navy recruiting office, and went back to his room to pack up. When his friends came in and found out what he was doing, they made him promise to try another major before dropping out. He picked up the course catalog, flipped through it, saw horticulture, and without knowing anything about it decided it would be his new major. After his first two courses, he knew he'd found his calling.

When we arrive at the half-acre of coleus, I'm stunned. The bold Florida sunshine lights up hundreds of waist-high mounds of the most kaleidoscopically colorful plants I've ever seen. These are nothing like the funereal ground-huggers I knew. These have leaves with hot pink centers and purple borders, leaves with a center oval of raspberry ringed first by vanilla and then lime green, chartreuse leaves deeply serrated, velvety purple leaves with grass-green frills, rust-colored leaves with orange hems, heart-shaped leaves of purest magenta, pale-green leaves with bloodred spatters, red leaves in a big butterfly shape with brilliant pink veins, and leaves that look like a medley of fruit sorbets. Looking at the coleus rows is like looking at one of those brilliant Matisse interior paintings where blue-and-white checkered tablecloth meets green-and-yellow flowered chair meets striped red-and-white wallpaper. Times one hundred.

I gape and exclaim. Dave beams, and starts pointing out his favorites. It's clear his heart is in coleus.

"You know," he says sheepishly, "I worked the first eight or ten years of my career at the university building this molecular biology program. Then one day I had a student come along who didn't want to do molecular biology; she wanted to do conventional breeding. So we started investigating crops, looking for the ones that had the most genetic variability, and came up with coleus. That was almost a decade ago, and every year our coleus program, which is just simple old-fashioned breeding, gets bigger and bigger."

Dave has revolutionized coleus in color, pattern, and size. I wasn't wrong that the coleus of my memory were small and gloomy. Until Dave got involved in the crop, they were. Coleus have traditionally been propagated by seed. In order to produce seeds, the plants had to make flowers. In order to make flowers, they had to draw down energy stored in their leaves. As soon as they began their reproductive phase, they stopped growing, and their lower leaves dropped off. By the late summer, they were leggy and unattractive.

Dave began breeding coleus with the goal of creating

plants that either never flower or if they do, flower sparsely and late in the season. He wanted plants that would channel their energy away from making reproductive organs and seeds and into making deeply pigmented leaves. Any wild coleus individuals that make that trade-off are goners, but coleus that do so in his garden are winners. What Dave has found through years of selection is that the red-blue anthocyanin pigments that protect leaves from ultraviolet rays can produce a rainbow of colors, especially if plants are not sidetracked into making babies. Plus, plants that never hit puberty continue to flourish and grow into the fall, sometimes reaching six feet high. Of course, all of Dave's plants must be reproduced vegetatively by rooting cuttings, which makes them six times more expensive than traditional varieties. Nonetheless, major growers—Ball, Proven Winners, and Syngenta—now license and sell Dave's varieties, and gardeners happily pay for long-lived, brilliant, and flowerless coleus.

PART V

Onward, Upward, and Afterward

twenty-seven

Trouble in Paradise

When I left Charles and Susan Farmer's citrus nursery, Charles promised to keep me up to date on my cocktail tree's progress. We spoke occasionally over the next months, and from time to time he sent me cell phone photos so I could see the branches growing longer and sturdier. After a year, however, USDA still hadn't implemented the regulations that would allow him to ship my tree.* I began to feel guilty. The Farmers are in the citrus and blueberry, not the plant-sitting, business. Then Charles mentioned that

---

* In 2013, a handful of larger companies were able to satisfy USDA's requirements and started to ship trees out of state. A treatment for the greening may be on the horizon. Dr. Erik Mirkov, a plant pathologist at Texas A&M University, has transferred genes from spinach into citrus trees, and the spinach genes appear to confer a resistance to the disease. These transgenic trees (including Hamlins) are in field testing in Florida. Still, the disease may well mean, as *Scientific American* put it, "the end of orange juice."

someone else, a Floridian, had called to ask about buying a cocktail tree, and I added worry to guilt. This tree was not going to slip out of my grasp.

I couldn't take it out of Florida, at least not without risking arrest, but I could move it within the state. My mother, eighty-four years old, widowed, and living in Fort Myers, became my Plan B. If she would take custody of my tree, at least I could enjoy it on my periodic visits. Perhaps there would be a time when I would be able to take it north.

I wasn't at all sure, though, that she would be amenable to my plan; gardening was of no more interest to her than it had ever been. If I had suggested she take care of an ordinary orange tree, I suspect she would have declined. But with this cleverly engineered tree, I thought I might have a chance.

As a high school student in the 1940s, my mother had been the only girl to take "shop." She is as manually gifted as I am not. As a young married woman with two toddlers, she was one of the original phone hackers: When AT&T was a monopoly and charged an additional fee for each "extension" in a house, my mother figured out how to splice into the wiring. She delighted in foiling Ma Bell, and we had a telephone in every room, including the little storage closet under the basement stairs. Together, my parents started a furniture-making company in our basement when I was about five. Where all the neighboring families had a rec room with a TV, a wet bar, and maybe a Ping-Pong or pool table, we had a woodworking shop with a table saw, stacks of lumber, and bags of cement. Their product was a line of poured-concrete coffee tables with insets of slate rectangles. (I suppose the tables had a Bauhaus sort of aesthetic.) They also made knock-down beds and knock-down sofas out of plywood, long before IKEA did. Their hands-on approach extended beyond furniture. My mother sewed many of my clothes when I was a child. Together they laid tile, and built an elaborate fence around the patio, a screened-in porch, and a wall of louvered cabinets. No professional painter, plumber, or electrician passed through our doors.

My mother assumed everyone was as dexterous as she, and organized an ice-carving birthday party for me when I was twelve. She set up sawhorse tables in the basement and every guest got a twenty-five-pound block of ice. We had our choice from a selection of tools from the furniture shop, mainly screwdrivers, but also awls, hammers, and paint scrapers. (Amazingly enough, no one was injured.) She also taught me how to use a sewing machine, and gave me lined notebook paper so I could practice stitching up and down straight lines. When I managed to do this, she signed me up for sewing classes at a local fabric store. The chief result was an off-white, corduroy culotte-dress that hung in my closet unfinished for years. I had enjoyed choosing the fabric and cutting out the pattern, and managed to sew the long seams. But the fine details—the interfacing, the facings, buttonholes, and hem—were too frustrating, and I returned, as my father said, to writing "pomes."

My parents were pragmatic, problem-solving people. I suspect they considered gardening an effete activity, sort of like poetry for the outdoors. Still, the yard couldn't go untended. My mother turned the side yard, which was shaded by the house, into what she called a rock garden. The job was half done before she started: The builder had dumped excavated material and some construction rubble there and then covered it up with fill dirt. The soil was just deep enough to support pachysandra and hostas. The sloping backyard she let grow into a scrubby thicket dominated by an untamed pussy willow tree whose lovely, soft, gray "kittens" appeared in early spring. Ahead of her time again, she maintained that she was creating a haven for birds, although I believe the thicket was really cover for her aversion to taking care of plants.

When I was thirteen, my parents fell in love with sailing, and bought a fourteen-foot Sunfish. My mother abandoned the petunias, the thicket, and the hostas, and every weekend our family drove to the Chesapeake Bay in a red VW Beetle, towing the Sunfish behind us. All four of us crowded onto this craft not much more substantial than a windsurfer. My younger sister Joanie and I vied for the narrow spot in front of the mast where we wouldn't get whacked in the chest by the boom.

Joanie and I were not enthralled with our parents' hobby, especially as we grew older. To us, sailing meant bobbing about in the crisscrossing motorboats' wakes on hot and windless days while the sail flapped and snapped. "Stinkpots," my father would scoff at passing Boston Whalers, but I admired those bare-chested teenage boys and the girls in bikinis with their hair streaming out behind them as they passed us. A cabin cruiser looked good to me, too. Inside, I figured there was air-conditioning, and wondered if the ride would be smooth enough so I could read. My melanin-deficient complexion condemned me, in those days before effective sunscreens, to a long-sleeved shirt and a hat; otherwise I'd be blistered by bedtime. After June, even the relief of swimming was out because the jellyfish appeared. Our baking boredom was interrupted only by occasional bouts of shivering terror when a sudden thunderstorm blasted through, bringing stinging rain and bristling with lightning that I was sure would strike the metal mast and fry us like soft-shell crabs.

Some time in our mid-teens, when my parents graduated to a nineteen-foot Mariner, a weekender with a cabin and a primitive toilet, they agreed we could stay home, and we became happy sailing orphans. Our parents became ever more committed sailors: over the next twenty-five years, they traded up again and again for an ever larger boat. As soon as my father was able to retire, he and my mother put their house up for sale, sold the furniture, put a few boxes in my attic, and headed south on the Intracoastal Waterway in a thirty-eight-foot Endeavour. They traveled slowly and contentedly around Florida and through the Caribbean, adding to their birding life-list and dropping in on local bridge clubs. Not until their mid-seventies did my mother's arthritis finally end their peregrinations, and they moved to a condo—with no yard—in Fort Myers.

All of which is to say, I didn't know if my mother's curiosity about a cleverly constructed cocktail tree would trump her lack of interest in plant care. To my relief, it did. She agreed to play host, but with two conditions. She would bear no responsibility for the outcome of her caretaking, and I would have to tell her exactly what to do for the tree, right down to when to water.

That last requirement should have scotched the deal, given

that I live a thousand miles away, but we—me, my mother, and the tree—were in luck. I'd already given a lot of thought to the challenges of proper watering. I'd been taking care of more than fifty indoor plants for nearly a decade, and still lost plants to dehydration. (Would that I were so mindful that my plants might run the risk of overwatering.) I know I should make a circuit of my menagerie first thing each day, lifting the pots or testing the soil with my finger for dryness. But watering is a chore, and I'm a procrastinator of the highest order, so I always figure I can make the circuit a little later. I mean, what difference could another hour make? Somehow, days go by before I finally notice that the leaves on the coffee are limp and my prayer plant is prostrate, not with piety but deprivation.

I had already determined that this would not happen to my cocktail tree, and had made preparations. I knew I couldn't reform myself, so I decided to enlist technology. In my roamings around gardening sites on the Internet, I had come across an ad for an electronic moisture meter, a do-it-yourself kit called Botanicalls, available for purchase online. Assemble the meter, plug it in, stick it into the soil of a potted plant, and as the soil dries, so its inventors claimed, the device would text or tweet you to let you know. I fell for it, cable, cord, and router, and ordered the kit.

When I opened the Botanicalls' box, I found a bright green circuit board in the shape of a saucy, ovoid leaf; two slender metal probes for detecting moisture; and a plastic bag with several dozen colorful transistors, resistors, capacitors, and LEDs. I had no hope of putting the thing together myself, but I am not without resources. Those resources are my cousin, Danny, who has a doctorate in computer science, and his clever thirteen-year-old daughter, Rosie. Danny and Rosie agreed to solder the components to complete the circuit board. A week later, they brought by the assembled device. I slipped the probes, now attached to the assembled circuit board, into the soil of a parched peace lily, and plugged the cord into a wall socket and the cable into a wireless router. A minute later, the plant sent a message to my cell phone. It arrived on my Twitter account.

"URGENT! Water me!" it read.

I poured in a cup of water.

"Thank you for watering me!" it responded. Over the next weeks, the peace lily sent me regular updates on the moisture content of the soil:

"Current moisture: 84%."

"Current moisture: 72%."

"Current moisture: 24%. Need water."

I added a cup of water, but this time received "You didn't water me enough!"

The Botanicalls turned out to be a terrific nag; the only way to make it stop was to water. I loved it. Danny taught me how to reprogram the device's script. The possibilities were limitless.

On my next trip to Florida, I drove up to Auburndale to meet Charles and pick up my tree. I was pleased to see that it now looked as full and vibrantly healthy as all the other Hamlins in the greenhouse. Each scaffold limb of my tree had a white plastic tag with its species name, and I could see that the shape of the leaves varied from branch to branch. The foliage of the variegated Eureka lemon had lovely cream-colored edges, like a ribbon around a hemline. Charles hefted the tree into my car and gave me last bits of advice and good wishes, and I drove her two hours south to Fort Myers.

I say "her" because I had decided my tree needed a name. I christened her "Dorothy" for Dorothy Parker, a woman who, like my mother, appreciated a stiff drink and might have gotten a kick out of a cocktail tree. (As Parker wrote: "I like to have a martini. / Two at the very most. / After three I'm under the table, / After four I'm under the host.") Mrs. Parker had been a scrappy soul, and survived three suicide attempts, a fact I thought boded well for my tree.

With the help of a neighbor, I settled Dorothy onto my mother's screened porch, plugged in the Botanicalls, set up a Twitter account for my mother, and signed her up to receive Dorothy's tweets.

On Monday, Tuesday, and Wednesday, Dorothy tweeted "Everything is copacetic." Thursday, she warned, "Feeling a little thirsty. Make sure we have tonic and limes." On Friday, the message was "It's almost five o'clock. Do we have ice?" On Saturday, I left for home, but my mother called later to tell me Dorothy had advised her to get out the glasses and the crackers and dip. On Monday, Dorothy tweeted: "Absolutely parched. I'll have a double." My mother complied, and poured in two quarts of water (straight up) until, per my instructions, water ran out of the holes in the bottom of the pot.

Dorothy wrote back, "Cheers!"

Two months later, I visited again, bringing my laptop and an FTDI cable for downloading a new script. All went well, with Dorothy reporting to my mother, and my mother reporting to me. (My mother, who has been known to pour an off-brand vodka into a Stolichnaya bottle, was nonetheless a little miffed when Dorothy warned, "And none of that no-name stuff.") In January, however, my mother reported that Dorothy was having a problem. Not a drinking problem, of that she was certain. Nor was Dorothy undernourished: My mother had added fertilizer on schedule. Nonetheless, the tree's lower leaves were turning yellow and falling off.

Shades of Kam Kwat.

I had my mother inspect the leaves for bugs. She didn't see any, she said, but there were some little, wispy cobwebs. And the leaves seemed oddly dusty.

I knew that syndrome. Dorothy was infested with spider mites. Spider mites are so small that to the naked eye they look like a fine, reddish dust. When an infestation gets bad, tiny cobwebs appear at the intersections of leaf and stem. Like other arachnids, spider mites spin webs, although not to trap prey but to avoid predation. Like aphids, they drill their stylets into leaves—into individual cells, in this case—in order to drain the sucrose, a practice that eventually kills the host. Outdoors, mites are controlled by predator insects and washed off by rain. On my mother's screened veranda, they were living in a gated, all-inclusive resort.

Fortunately, the population of spider mites on a single plant is easy to reduce, although it's hard to eliminate them all. I told my mother that she'd have to spray the leaves, particularly their undersides, with horticultural oil once a week for three weeks. The first spraying would smother the active bugs; the next two would take care of the hatchlings of eggs already laid. I expected some resistance to that chore, but to the contrary, she was eager to get started. Of greater concern was how she could get new leaves to come out on the defoliated branches. Dorothy, she said, was looking rather threadbare. I promised I would figure it out. At least I knew radical pruning was not an option. To my surprise, I discovered that Charles Darwin had something to do with the solution to Dorothy's problem.

twenty-eight

Onward and Upward

In the early 1860s, at the same time Darwin was observing orchids and pondering how evolution shaped their extraordinary variety, he became infatuated with the sundews (*Drosera*) that grew along the edges of a nearby heath. Sundews are small plants that live in boggy areas in North America and northern Europe. Their leaves, round or straplike, are rimmed by sticky filaments that snag insects unfortunate enough to blunder into them. The leaves then slowly bend over the victim and dissolve it. Darwin was fascinated by the process and set out to discover what the *Drosera* eat. He fed his sundews raw beef, peas, olive oil, and even cobra venom, all of which they readily devoured. When sand, cinders, and glass were on the menu, however, although they closed their leaves over these tidbits, they soon reopened. Nitrogen, he concluded, is what sundews seek, and their carnivorous ways are an evolutionary adaptation that allows them to live on nitrogen-poor soils. In fact, we now know that up to 50 percent of a sundew's nitrogen comes from digesting trapped insects.

In 1863, Darwin began to study climbing species to explain how evolution had enabled plants without trunks or stiff stalks to join in the general race toward the sun. To the congeries of exotic orchids and sundews in his heated greenhouses, he added climbers. Plants with weak cores have three strategies for using external supports to compensate for their deficiency. Some, like hops, are twiners that travel sunward by twirling their stalks around a buttress. Others have projecting tendrils or leaves (the pea and potato plant, respectively) that do the grasping. But how does the tip of a stem or tendril or leaf find something to grip? He could see the tips moved, seemingly in search of a handhold, but was there a method to their investigation?

By positioning a pane of glass above these moving extremities, Darwin was able to trace their paths throughout the course of a day. He discovered that they rotate in an elliptical, species-specific pattern, which he called *circumnutation* or "nodding in a circle." When a tip encounters a vertical object such as another plant's stalk or a fence post, by continuing to circle, it gradually bends itself around the rigid structure. Because the tips grow upward while they circumnutate, they tend to form a spiral. Darwin had noticed that when a tip encounters a crevice in a flat surface, it appeared to consciously explore and then reject the location, but he now saw that actually it is pulled out by its own constant revolving and upward motion. In *The Movements and Hab-*

its of Climbing Plants (1865), he concluded that the various strategies for clinging likely evolved from some ancient capability. Twining, he wrote, is another example of modification by descent.

Darwin next returned to carnivorous plants, and in particular to the Venus flytrap, a North American native he called "one of the most wonderful plants in the world." At the top ends of the plant's leaves are a pair of large, reddish lobes joined in the middle and rimmed with needlelike tines. When the trap snaps closed—in about a tenth of a second—the tines interlock to create a barred cage.

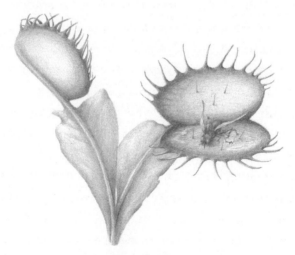

Venus flytrap.

Flytraps, he recognized, have the same gustatory preferences as sundews, but have evolved more effective hunting methods. In addition to nectar on their surfaces and tines around their edges, each lobe has three touch-sensitive hairs, called trigger hairs, on the interior surface of each lobe. A trap closes only if two of the trigger hairs are touched within twenty seconds. In this way, unlike the sundews, flytraps don't waste energy closing over a cinder or a raindrop that is likely to touch only a single filament. Moreover, the tines of the flytrap's cage are spaced so that smallest insects can escape; the amount of nitrogen they contain isn't worth the metabolic cost of digesting them. The more refined mechanisms of the flytraps, Darwin concluded, had evolved from the simpler sundews, and provided additional evidence for his theories.

In the mid-1870s, Darwin returned to studying plant movements, finding them interesting in their own right, and set out to understand the physical and chemical mechanisms that activated them. In essence, in his late sixties, he decided to branch into a new scientific discipline, plant physiology. Unlike his previous work, which was based primarily on his exquisite skills as an observer—as well, of course, as his powerful and creative intellect—his new work would involve experimentation. He started by identifying what he considered the most basic kind of plant movement, the one that climbing plants must have modified over time. That movement, he concluded, is phototropism, a plant's ability to bend to follow the sun as it moves across the sky over the course of a day. Phototropism (or as Darwin called it, heliotropism) would be the subject of his experiments.

In 1877, however, he was sixty-eight years old, and wasn't sure he had the energy to undertake the work. In the eighteen years since the publication of the *Origin,* he had revised his great opus six times, rewriting about 75 percent of the text to incorporate new evidence. In the same period, he had researched and written articles and books on subjects as diverse as human evolution, flowers and fertilization, barnacles, orchids, climbing plants, and insectivorous plants, as well as engaging in a massive and international correspondence. All of this intellectual activity took place at his home, where he also participated in the lives of his wife, Emma, and their seven surviving children. Moreover, throughout these intense years of creativity, and in fact ever since his return from the *Beagle,* he had battled bouts of depression and episodes of unexplained vomiting, vertigo, tremors, and other symptoms that had interrupted his work, sometimes for months at a time. Now the symptoms were worsening.

To make matters much worse, his son Francis had recently lost his young wife to puerperal fever, leaving Francis the devastated father of a newborn son, his first child. Francis and his wife Amy had lived just a quarter mile down the road from Charles and Emma, who had known Amy's family for years. Amy's death was not only a present loss—Charles relished her high-spirited, outgoing nature, so unlike his own, and treated her like a daughter—but it opened an old wound that had never fully healed, the death of his firstborn daughter, Annie.

Annie had had a similarly warm and demonstrative personality, and Darwin admitted she was his favorite. When Annie was eight, she contracted scarlet fever, and her health was never the same. Two years later, Charles had taken her to the spa at Malvern, 150 miles northwest of London, where he had often retreated to tend to his own ills. Shortly after their arrival, Annie developed a high fever, and died. In a memorial he wrote at the time, Darwin mourned her "buoyant joyousness" and the depth of her affection, which made her exquisitely sensitive to the moods of her parents. Charles also suffered with an extra burden of guilt. He worried that he had passed on his own poor health, and, moreover, that her frailty was compounded by what he had come to understand were the dangers of inbreeding. Because Emma was his cousin, he feared he had inadvertently doomed his beloved child.

In an effort to distract Francis from his grief and in great need of distraction himself, Charles proposed to his son that they engage in a joint research project. Together, they would look for the answer to how it is that stems, without any muscle tissue, manage to bend toward the sun.

Francis agreed. He had earned his B.A. from Cambridge in 1870 and a medical degree from the University of London in 1875, and had seemed headed toward a career in medical research, with three significant research papers on animal physiology to his name. Nonetheless, after Amy's death in 1876, father and son formed a professional partnership. Charles was accustomed to creating controlled conditions in his greenhouse and collecting data, but he was not a laboratory scientist. Francis was newly trained in the latest lab techniques and design of experiments.

Francis's experience is clear in the approach the pair took in investigating phototropism. The Darwins' experimental subjects were the emerging seedlings of Canary grass and oats. After germinating seedlings in darkness, they illuminated them from one side only. The seedlings curved toward the light. To determine what part of the seedling could "see," they wrapped the tiny stems, except for their tips, in an opaque cloth. The seedlings still bent toward light. Next, they excised the seedlings' tips, and found that the stems were blinded and continued to grow upright. To eliminate the possibility that the stems failed to curve, not because the tips were missing but because the seedlings had been damaged, they fashioned little caps from feather quills, blackened them with ink, and set them on the seedlings. Again, the seedlings failed to respond to light. When they removed the quill caps, the little plants curved once again. The Darwins concluded that some chemical agent in the tip transmits an "influence" that causes the stem to curve.

What was this influence? Scientists of the period didn't have the tools or the chemical knowledge to find out. But in 1926, the Dutch botanist Frits Went ran a most elegant iteration of the Darwins' experiment. Went cut off the tips of grass seedlings and rested them on a sheet of gelatin, so that any liquid in the tips would drain into the gelatin. Sometime later, he cut the gelatin into tiny cubes, and placed one cube on the top of each decapitated stem, positioned so that it covered only one-half of a stem. He found that the half of the stem beneath a cube grew more than the uncovered half, inducing a curve. Moreover, the greater the number of seedling tips he placed on the gelatin before carving it into cubes, the more sharply any stem would curve after being partially capped. In other words, the greater the concentration of this mystery substance, the more curvature it induced. Went was able to isolate the chemical and named it *auxin,* from the Greek *auxein,* meaning to grow or increase.

Meristematic tissue in stem tips produce auxin. When

light hits the meristem from one direction, auxin migrates away from the light and toward the opposite side of the stem tip. The extra auxin causes those cells on the far side to elongate. As the extra auxin flows down through the cells along the far side of the stem, they also elongate. The result is a stem that gracefully arcs toward light. As the sun moves across the sky, different cells are flooded with auxin, causing the stem to follow the source of light.

Auxin was later identified as one of five major types of chemical messengers, *phytohormones,* manufactured by plants. Phytohormones control gene expression, germination, plant development, the time of flowering and fruiting, leaf drop, the opening and closing of stomata, senescence, and other activities. Unlike animal hormones, which are produced in glands, phytohormones are produced in a plant's cells. Without them, a plant would look like a blob of algae.

In modern tissue culture labs, technicians use phytohormones to turn a few cells from a parent plant into thousands of identical seedlings in a few weeks. By applying various synthetic hormones in exact amounts at precise intervals, the technicians completely control cell multiplication and root and shoot development. The seedlings, often barely an inch long, are then sold to wholesale nurseries, where growers pot them, add more hormones to make them grow bushy, and then later add more hormones to make them flower synchronously. One of the reasons houseplants often fail to thrive after I bring them home from the garden center is that I don't have access to synthetic phytohormones. On the other hand, if I were to use a weed killer like Verdone on my lawn, I would be deploying phytohormones to ensure that certain weeds *don't* thrive. These herbicides contain a synthetic auxin that spurs dandelions, chickweed, clover, and other broad-leaf plants to grow so fast that their stems twist and contort, and the plant dies.

So, what do Darwin's discoveries have to do with Dorothy and her dropped leaves?

Trees have evolved to hoist their leaves above their competitors. In order to concentrate their energy on growing taller, they have developed a phenomenon called *apical dominance,* which means the stem or stalk tips or *apices* (the plural of apex) grow more strongly than shoots emerging at the sides. Apical dominance is primarily a function of auxin. While auxin causes cells in the apical meristem to divide and expand, it also influences other plant tissues. Specifically, as it travels down the stem, it has the *opposite* effect on meristematic tissues farther down the trunk, stalk, or stem. Auxin prevents axillary buds, which sit atop a dome of meristematic tissue, from opening and slows the growth of lateral branches.

In some conifers, the effect of apical dominance is dramatic. Auxin produced in the apical meristem of a fir tree trunk diffuses downward and arrives in nearly full strength at the topmost branches, where it greatly inhibits their growth. A little less auxin makes it to the next lower branches, which grow a little longer than the ones above them. So it goes, down the trunk, resulting in a tree with a classic Christmas tree shape. Strawberry plants, on the other hand, have little auxin in their stem tips and therefore exhibit little apical dominance and hardly grow in height at all. Instead, they compete for sunlight by spreading out along the ground. I had been right to prune my kumquat, thereby removing the apical meristems on its branch tips and their inhibiting auxin. My mistake was in my drastic application of the principle.

Deciduous trees, like the crape myrtle Ted successfully prunes so radically, have evolved to deal with surviving cold months. In winter, the energy that a tree's leaves are able to generate during short daylight hours is less than the energy required to maintain cell function in the leaves. In addition, the loss of water through transpiration exceeds the amount that the roots are able to absorb when groundwater is locked up in ice. So, in autumn, deciduous trees cut their losses. First, thanks to hormonal signals, they drain the sucrose from their leaves and send it to their roots and branches for storage. Then they seal off the leaves at their bases with a corky substance. Without water and nutrients, the leaves' cells die. In the spring, the trees send stored sugar dissolved in water up the xylem to fuel the growth of new leaves and branches. It is that rising sap in sugar maples that New Englanders tap to make syrup.

Tropical and subtropical evergreens like the kumquat and Dorothy, on the other hand, have evolved in regions where the number of daylight hours hardly varies throughout the year and temperatures rarely fall below freezing. These trees have less need to store carbohydrates in their roots. After I gave my kumquat a buzz cut, it didn't have enough stored energy to rebuild its leaves, and died.

Dr. Timothy Spann at the University of Florida's Citrus Research and Education Center told me how I should take care of Dorothy. There are two techniques, he said. I could prune one-third of her branches. Then, after the axillary buds on the pruned branches have opened and their new leaves fully matured, I could prune another third of the branches. After waiting again, I would prune the final branches. Or, for faster results, I could try tricking my tree. Faster is always better in my book, so I took notes.

My first step would be to grab any branch flexible enough that I could bend it into a ⌒-shape, so its tip pointed to the ground and an axillary bud was at the apex of the arc. Then I would tie the branch in that position. (If a branch was too stiff to bend, I could cut partway through its diameter—a technique known as "lopping"—before bending and tying the branch.) Because auxin at the branch tip cannot flow against gravity, it would not reach the axillary buds, which meant the lower buds would be uninhibited and free to grow. Meanwhile, the upside down leaves would continue to photosynthesize and transpire, providing energy and pulling water and nutrients up from the soil.

So, on a pleasantly cool March morning when I was visiting in Fort Myers, I set to work on Dorothy with my mother in anxious attendance. First, as if I were prepping a patient for surgery, I trimmed off Dorothy's wicked thorns. (As Parker once said, "The first thing I do in the morning is brush my teeth and sharpen my tongue.") Then I doubled over a slender branch and held it in place while my mother tied the limb in place with a piece of green twine. We worked our way around the tree from bottom to top. Most of her branches were sufficiently supple to double over, and only a few needed lopping. It wasn't long before Dorothy looked like a thoroughly trussed turkey. I left for home, leaving my mother on duty.

Nothing happened for three weeks, and while the patient

remained perfectly calm, her nurse fretted daily. Then my mother called one morning to report that, just as Dr. Spann had promised, the axillary buds along the upright portions of the arched branches were opening. Dorothy, she said, was sporting pin feathers. My mother sent photos, and after a month, we agreed that it was time cut the twine and trim the upside-down branch ends. By August, when I visited again, Dorothy had regained her full plumage and even sported some purple flower buds.

My mother continued to send photos regularly—I felt like a devoted but distant grandparent—as the tree bloomed and began to fruit. At my Christmas visit, she proudly showed me Dorothy decked out in bright yellow Meyer and green-and-yellow-striped Eureka lemons, dark green limes, two orange Cara Caras, and a single Minneola. Even if the citrus regulations change, I think Dorothy has found her permanent home.

twenty-nine

Afterward

During the second half of the twentieth century, the focus of botanical research gradually shifted away from physiology and into molecular biology, which is the study of DNA, proteins, and other molecules involved in cell function. Now we are seeing a synthesis of those two disciplines, a synthesis spurred by advances in genomics (the identification and sequencing of genes) and genetic engineering. Bioengineers' ability to transfer genes into and out of chromosomes in order to change the functionality of plant tissues and organs is opening new worlds in botany. Should we convince ourselves of the safety of genetically engineered plants or that any risks are outweighed by benefits, there will be lifesaving—and Earth-saving—opportunities in store.

For example, more than a decade ago, Igor Potrykus and Peter Beyer, plant scientists working in Switzerland and Germany, respectively, transferred three genes—two from a daffodil and one from a bacterium—into the genome of rice. The added genes express a group of molecules that when ingested by humans becomes vitamin A. They also impart a yellow hue to the rice, giving it its popular name, "golden rice." Golden rice has the potential to correct vitamin A deficiency, a health problem that, according to the World Health Organization, harms 250 million preschool children in developing countries and blinds as many as 500,000 each year. The rights to the rice are now held by the Golden Rice Humanitarian Board, and any farmer in a developing country who makes less than ten thousand dollars a year can, hypothetically, get free seeds. However, the high cost of meeting regulatory requirements and concerns about whether consumers in poor countries will accept a nonwhite rice means golden rice as yet grows only in test fields.

Abscisic acid (ABA) is a phytohormone that helps plants cope with the stress of dehydration. When drought strikes, the hormone turns on receptors that cause stomata to close, and slows or stops growth so a plant needs less water. In 2009, Dr. Sean Cutler and his colleagues at the University of California, Riverside succeeded in engineering the ABA genes of *Arabidopsis thaliana,* another white mouse of botany labs, so that they can be turned on at will. Their work may lead to crops better able to survive and produce in regions where drought already limits—or will soon limit—harvests.

Sixty percent of the world's population lives in Asia, and many Asians depend on rice to survive. An average hectare of land in Asia devoted to rice, according to the International Rice Research Institute (IRRI), feeds twenty-seven people. By 2050, population growth will mean that each hectare will need to feed forty-three people. Relying on increases in rice yield to make up the difference is problematic. The annual rate of yield increases has fallen by half since the 1990s, as efficient farmers approach the fundamental limits of how much sunlight rice, which uses C3 photosynthesis, can convert to sugars. In fact, the productivity of rice plants will likely decline because the C3 pathway becomes less efficient the warmer the climate becomes.

To raise the theoretical limit on rice productivity, IRRI scientists, funded in part by the Bill & Melinda Gates Foundation, are working to develop varieties that use the more productive C4 pathway. With traditional breeding techniques, they are trying to activate certain C4 cell structures and enzymes currently quiescent in the rice genome. Not every capability is available, however, and they may need to transfer genes from species like maize and sugarcane. IRRI estimates that a C4 rice could produce about 50 percent more carbohydrates than today's varieties.

Eukaryotic algae manufacture oils as well as sugars that they use to grow and function. Some algal species are theoretically capable of producing more oil than oilseed crops like sunflower and soy. That oil can be used for transportation fuel. A number of companies, including joint ventures involving major U.S. oil companies, are making progress in producing algae oil for human energy use, and a few airlines and the American military have incorporated it in jet fuel. Unlike farming corn to make ethanol, growing an algae-oil crop doesn't use up arable land that would otherwise be used for growing food. (Even better, while growing corn requires large quantities of freshwater, many algal species thrive in brackish or salt water, of which the Earth has no shortage.)

However, making the fuel at a large scale and in a profitable way is still a challenge. For one, getting the oil out of algae is surprisingly difficult and expensive.

Perhaps a more promising path lies in genetically engineering cyanobacteria. The genomes of cyanobacteria, like that of other prokaryotes, are organized in a simple loop rather than the helices, which means it is relatively easy to add foreign DNA to their genomes. Bioengineers are genetically modifying cyanobacteria so that they produce ethanol, butanol, alkanes, and other fuels. Even better, the cyanobacteria secrete rather than store these products, so collecting them is far easier. No one has yet to prove the commercial viability of cyanobacteria fuel, but companies like Algenol and Joule Unlimited and others are betting hundreds of millions of dollars that it can be done.

It doesn't take such advanced techniques, fortunately, to put the science of plants to work in the garden. Those two, tree-size dracaenas I have, whose roots had snaked out of the drainage holes of the pots to circle the saucer? Once I understood that woody roots have little to do with absorbing nutrients and water, I clipped them off, and then repotted the plants in Air-Pots. A British invention, an Air-Pot looks as though it has been formed from a black plastic egg carton. On the inside of the pot are inward-pointing, closed cones. When a root touches one of these cones, its growth is directed toward a proximate, outward-facing cone. The outward-facing cones have holes, and when a root tip grows through the hole and hits the air, it desiccates and dies. This "air-pruning" kills the apical meristem of the root tip and, in a manner similar to pruning branch ends, stimulates new lateral roots to sprout. More lateral roots mean more root tips that access more water and nutrients. After I repotted my rootbound dracaenas in Air-Pots, they stopped shedding leaves, and now look healthier. Better living through plant physiology.

I recently ordered a new kumquat tree, a Meiwa, billed as the easiest kumquat to grow and the sweetest to taste. It came, bare-rooted, from California, and I planted it in a five-gallon Air-Pot. Of course, I added mycorrhizae to the potting mixture, as well as a slow-release fertilizer with all the essential macro- and micro-nutrients. A pH meter tells me if the soil is in the range where minerals can be effectively passed by membrane transporters toward the xylem. I no longer wait for leaves to wilt as the sign for me to water. When many of the pea-sized, green kumquats drop from the tree in spring, I don't worry: It's just the tree balancing the energy it spends on building fruit versus its other metabolic needs. I'm quick to prune the branches (no more than a third at a time) to avoid that gangly look. I'm quicker to attack the spider mites, especially when my tree is indoors, where marauding mites are safe from natural predators. I used to think the mites were just an unsightly nuisance, but I understand now. Their thefts, unchecked, mean the tree has fewer resources to repair damaged cells or build new ones. My new kumquat is doing well, thank you, and produced dozens of kumquats last winter, just when my spirits needed a boost.

So, I can say that I am a better, less lethal gardener for having begun the study of botany. Now at least most of my mistakes are due to inattention, and fewer are the result of ignorance. But best of all, I have had the pleasure of an invigorating intellectual journey. I now see night-blooming jasmine in a new way: Its highly scented, white flowers are appropriate attire and perfume for a hookup with a nocturnal moth. I understand why I've never noticed the flowers of oak trees: Why would they spend their hard-earned savings on high fashion when no one comes to call? I'm tickled to know how pollen tubes drill through a flower's pistil, awed by the complexity of flower color, and not surprised to find a yellow-striped bloom on my Black Velvet petunia. But the greatest marvel of the garden, I feel, lies in leaves. Gaze at a single blade of grass and you are witnessing tens of millions of microscopic green engines, capturing photons, splitting water, and manufacturing sugars with carbon snatched from air. These engines create 99 percent of the biomass on Earth; all of us oxygen-breathers constitute just 1 percent. Every day, in the improbably complex process of photosynthesis, chloroplasts turn hundreds of millions of tons of carbon dioxide into oils, carbohydrates, and, by incorporating nitrogen and sometimes sulfur, proteins. Simultaneously, they release oxygen, almost all of which is used up by respiring humans and other animals, fungi, and bacteria.

Who hasn't looked at the stars in the night's black sky and been humbled by their own smallness and insignificance? But now when I look out over my leafy neighborhood from the window of my third-floor office, I think of this. There are vastly more chloroplasts on Earth than stars in the universe. All these chloroplasts owe their lives to that one eukaryote that engulfed an indigestible cyanobacterium that lived 1.6 billion years ago. That single creature's descendants turned the rocky continents into our leafy, green world, without which none of us could exist. Our garden is more than a marvel. It's as close to a miracle as there is on Earth.

Acknowledgments

I had so much help in writing this book. Thanks to all of the individuals who appear in these pages who spoke with me about the subject of their research, their hobbies, and their businesses, and gave me the benefit of their expertise. In addition, I owe thanks to Dr. Pam Soltis, Dr. William Castle, and Dr. Loren Weiseberg for their insights.

Eva Ruhl produced these spectacular drawings with, what seemed to me, magical ease. She lives down the street, but at the opposite side of the bell curve of dexterity from where I dwell. We have had a wonderful time exploring together the way plants work and the way they look. I admire her work and her insights tremendously.

I have had the privilege of working with Jennifer Brehl twice now, and her enthusiasm for this book has meant much to me. I am grateful to Emily Krump and Rebecca Lucash at William Morrow, and to Tom Pitoniak for copy editing. Michelle Tessler has been my stalwart advocate and an excellent advisor. I thank Dr. Thomas Colquhoun at the University of Florida for bringing his scientific eye to bear on the manuscript and Alan Bateman for help on calculations; whatever errors remain, however, are mine alone. Kenny Greif, my mentor of more than forty years, has once again given me invaluable editorial input. Ellen Roberts has given me the benefit of her highly honed editorial skills. Thanks to my mother, Alice Good, for being such a good sport about appearing in these pages, and to Jim Howard and Kate Mendeloff for their reminiscences. My daughters, Anna, Austen, and Alice, have been a great support in this long endeavor, and I couldn't have written about Botanicalls without Danny and Rosie Edelson. Without Ted's encouragement, this book certainly would not have been written.

Notes and Sources

For the history of plant physiology, Morton's *History of Botanical Science* was essential reading for me, and his early chapters can be read without any knowledge of botany. Later chapters require Botany 101. Morton's work built on Julius von Sachs's two-volume *History of Botany,* but keep in mind Sachs is biased in favor of German scientists. *Nature's Second Kingdom,* by Francois Delaporte, was also useful as historical background, especially on the tradition of understanding plants by analogizing from animals.

For the basic science of plants, I relied on Linda R. Berg's *Introductory Botany,* Martin Rowland's *Biology,* and Peter Scott's *Physiology and Behaviour of Plants.* A useful primer for botany is Brian Capon's *Botany for Gardeners.* I also recommend khanacademy.org for its blackboard lectures on photosynthesis.

PART I: INSIDE A PLANT

The Birth and Long Life of the Vegetable Lamb

Henry Lee's book *The Vegetable Lamb of Tartary; A Curious Fable of the Cotton Plant* (1887) collects the stories about the borametz. Robert Carrubba's article "Englebert Kaempfer and the Myth of the Scythian Lamb" is also interesting.

Erasmus Darwin, grandfather of Charles and a poet and a botanist, wrote fancifully of the borametz in "The Botanic Garden" (1791), a two-part poem he hoped would "induce the ingenious to cultivate the knowledge of Botany":

> *E'en round the Pole the flames of love aspire,*
> *And the icy bosoms feel the secret fire,*
> *Cradled in snow, and fanned by Arctic air,*
> *Shines, gentle Borametz, thy golden hair;*
> *Rooted in earth, each cloven foot descends,*
> *And round and round her flexile neck she bends,*
> *Crops the grey coral moss, and hoary thyme,*
> *Or laps with rosy tongue the melting rime;*
> *Eyes with mute tenderness her distant dam,*
> *And seems to bleat—a "vegetable lamb."*

Darwin wrote a note on this stanza:

> *Polypodium Barometz.* . . . *This species of Fern is a native of China, with a decumbent root, thick, and every where covered with the most soft and dense wool, intensely yellow.* . . . *This curious stem is sometimes pushed out of the ground in its horizontal*

situation by some of the inferior branches of the root,
so as to give it some resemblance to a Lamb standing
on four legs. . . . The down is used in India externally
for stopping hemorrhages, and is called golden moss.

Despite the fact that England was importing huge quantities of raw Indian cotton for its textile industry, he failed to connect the borametz to *Gossypium* and the fluffy fibers that surround the seeds and help disperse them.

Through a Glass, However Darkly
Although Robert Hooke was not much interested in exploring the inner structure of plants, he was curious about what makes the nettle sting. He saw "sharp needles," which he found "by many tryals" to be "hollow from top to bottom." By putting a magnifying glass in a frame and attaching it to earpieces, he made magnifying eyeglasses and freed his hands to manipulate the nettle. Pushing one of its needles into his skin, he observed that his action caused the bottom of the needle to depress a small bag at its base. A fluid flowed out of the bag, up through the needle, and into his skin, and he correctly concluded that it is the fluid, not the prick, that causes the sting. We now know that the liquid contains a mix of neurotransmitters, including acetylcholine and histamines.

Hooke experimented on his own body in search of substances that would allow him a clearer, less depressed, more imaginative mind to handle all the work he had taken on. He systematically and almost daily swallowed substances, most of which we now know to be toxins, and noted their effects in his diary. Iron compounds, ammonium chloride, antimony, absinthe (which is made from various *Artemisia* species), and other "remedies" went down his throat at night, often on a tide of small beer or ale. Not surprisingly, he often vomited, had horrendous diarrhea, suffered cloudy vision, felt light-headed and feverish, and had numb extremities, but often also reported that he felt "refresht" or "strangely refresht" in the morning. As his body acclimated to the poisons, he needed to take larger doses to achieve an elevated mental state.

Hooke became involved in many rancorous disputes with his colleagues over who ought to get credit for inventions and discoveries. Most notably, he argued with Isaac Newton over who first expressed the theory of universal gravitation and elliptical orbits, and with Christian Huygens over the invention of the spring-regulated watch. (Good, but not definitive, arguments can be made for Hooke.) He had a rich social life into middle age, visiting and dining with friends and colleagues in their homes and coffeehouses and taverns. He had sexual relationships with several of his servants and his niece, but never married. In his later years, he alienated many of his former friends and colleagues, and it may well be, as Lisa Jardine suggests in *The Curious Life of Robert Hooke: The Man Who Measured London*, that the toxins to which he became addicted heightened his anxieties, his irritability, and his fears of being slighted. His emotional and physical suffering, as well as his self-medication experiments, no doubt hastened his death in 1703.

It is well worth looking at a copy of *Micrographia*. Not only are the drawings beautiful, but Hooke's text is quite readable and revealing of his searching mind. Bradbury provides the basic history of the microscope in *The Evolution of the Microscope.* He also passes along Henry Power's 1664 poem, "In commendation of Ye Microscope," which captures the self-consciousness of the early Enlightenment philosophers and their awe:

> *Of all the Invention none there is Surpasses*
> *The Noble Florentine's Dioptrick Glasses*
> *For what a better, fitter guift Could bee*
> *In this World's Aged Luciosity.*
> *To help our Blindnesse so as to devize*
> *A paire of new & Artificial eyes*
> *By whose augmenting power wee now see more*
> *Than all the world Has ever doun Before.*

Marian Fournier's *The Fabric of Life: Microscopy in the Seventeenth Century* explores the contributions of each of the important microscopists: Hooke, Malpighi, Grew, Leeuwenhoek, and Jan Swammerdam (who is best known for his pioneering anatomy of insects and for discovering red blood cells). *The Invisible World: Early Modern Philosophy and the Invention of the Microscope* is an eye-opening discussion of the significance of the instrument in the Enlightenment.

I had the good fortune to come across the collection of antique and replica microscopes at the University of Notre Dame, and Dr. Phillip Sloan helped me look through a number of early replicas. The experience gave me an even greater respect for the early microscopists: The tiny field of vision, the distortions, the imprecise focusing mechanisms, the awkward devices for holding specimens, and the lack of modern illumination would have been overwhelming obstacles to me. If South Bend is not in your travel plans, the catalog of the collection has photographs of the instruments.

Jardine's biography of Hooke is excellent, especially the discussions of his personality. (I buy her argument that Hooke was not as obnoxious as other biographers believe.) Stephen Inwood, in his superb *Forgotten Genius: The Biography of Robert Hooke,* details the mind-boggling breadth of Hooke's scientific investigations and accomplishments. "England's Leonardo: Robert Hooke (1635–1703) and the Art of Experiment in Restoration England," by Allan Chapman, provides another view of Hooke, the scientist.

For this and following sections, I relied on a number of sources on science, religion, and the Royal Society in the seventeenth century. The literature on the scientific revolution is extensive. Notable, for my purposes, are *Establishing the New Science: The Experience of the Early Royal Society; The University of Cambridge: A New History; The University of Cambridge and the English Revolution, 1625–1688; Revolutionizing the Sciences: European Knowledge and Its Ambitions, 1500–1700;* and *The Emergence of Science in Western Europe* (chapter 4). Marjorie Hope Nicolson's *Pepys' Diary and the New Science* gives an excellent flavor of what experimental science meant, if not to the average man, then to an educated and very curious layman.

The Persecuted Professor
The essays in the first volume of Howard B. Adelmann's magisterial, five-volume edition of Malpighi's correspondence are my primary source of biographical information about

the scientist. Adelmann is also highly informative about the history of Bologna, medicine, and Italian universities in the Middle Ages and the Renaissance. In addition, I consulted *Marcello Malpighi, Anatomist and Physician,* edited by Domenico Meli, and found chapter 9 of Dr. Meli's *Mechanism, Experiment, Disease* and "Mechanistic Pathology and Therapy in the Medical *Assayer* of Marcello Malpighi" especially helpful.

As Giovanni Ferrari points out in "Public Anatomy Lessons and the Carnival: The Anatomy Theatre of Bologna," the teaching of human anatomy originated in Bologna at the beginning of the fourteenth century. Although the anatomy lesson in Malpighi's era with its attending masqueraders was riotous by our standards, the event had actually become much more decorous and orderly than in previous centuries. In earlier times, students crowded around the table and jostled each other, trying to get their hands on the cadaver.

Malpighi's life improved in his later years, as his renown spread across Europe and the value of microscopy became increasingly obvious. (Sadly, the reputation of the university, once a preeminent center of learning, declined precipitously as Enlightenment science passed it by.) He continued anatomizing every living being he came across, providing the first understanding of the anatomy of uncounted species. The College of Doctors of Medicine finally inducted Malpighi into its ranks in 1690. (Professor Sbaraglia did not attend the ceremony.) The following year, the new installed Pope Innocent XII asked Malpighi to become his personal physician, and Malpighi, despite serious kidney disease, agreed. Three years later, at the age of sixty-six, he died of a stroke. His body, at his request, was autopsied, no doubt by men who had known him in life.

Inside a Plant

Unlike Malpighi, Grew wrote few letters and left few (figurative) bones for biographers to gnaw on. After his tenure as curator, he stepped into Oldenburg's shoes as secretary of the Society after the diplomat's death in 1677. In November 1679, he declined to continue in the job (leaving it to Hooke), and returned to the practice of medicine. According to William R. LeFanu, in "The Versatile Nehemiah Grew," Grew and his half brother, Henry Sampson, were admitted to the College of Physicians as Honorary Fellows, "a new rank intended to regularize the position of country physicians or holders of foreign degrees." Grew continued to contribute papers to the Royal Society on distilling seawater, skin ridges, the anthropology of the Indians of New England (based on questionnaires), snow, and a hummingbird. If you've ever taken a bath with Epsom salt to relieve muscle soreness, you should think of Grew: He was granted a patent for the product in 1698. (Epsom salt is magnesium sulfate, which, by the way, works by releasing magnesium ions that are absorbed by the skin and travel to the brain, where they interfere with pain receptors.) Toward the end of his life, Grew published a philosophical work that purported to prove the necessity of accepting Christian revelation. A kind and gentle man and a devoted physician, he died while making his medical rounds on March 25, 1712.

Malpighi wrote in Latin and Italian; Grew wrote primarily in English. *The Anatomy of Plants* was reprinted in 1965 in facsimile, and it is a pleasure to look at the drawings, which vividly demonstrate how Grew perceived plant tissue to be a woven material. The introduction in the reprint by Conway Zirkle is helpful, as are the articles on Grew by Jeanne Bolam and Agnes Robertson Arber.

Popular Medicine in Seventeenth Century England and Birken's article in *Medical History* were sources for information on the options for Grew and other Nonconformist medical practitioners. For background on the educational opportunities (and lack thereof) for Nonconformists, see Evans, Twigg, and the biographies of Priestley noted below.

PART II: ROOTS

Restless Roots

For discussions of the mechanical forces that keep trees upright, see Roland Ennos's essay "Trees: Magnificent Structures," on the Museum of Natural History's website (http://www.nhm.ac.uk/nature-online/life/plants-fungi/magnificent-trees/) and chapter 9 of Robert Kourik's *Roots Demystified*.

The Way of All Water

The best sources on the Great Chain of Being are Lovejoy's *The Great Chain of Being* and "The Great Chain of Being After Forty Years: An Appraisal," by William Bynum.

Most useful on Hales is *Stephen Hales: Scientist and Philanthropist* by D. G. C. Allan and Robert Schofield, which amplifies Clark-Kennedy's *Stephen Hales D.D., F.R.S.: An Eighteenth-Century Biography*.

How to Kill a Hickory

Roots Demystified by Robert Kourik is not only informative about the physiology of roots, but also oriented to gardeners. *Plant Roots: The Hidden Half* is a collection of essays: comprehensive but technical. To fully appreciate roots, you have to appreciate soil. I recommend *The Nature and Properties of Soil,* a popular textbook by Brady and Weil.

Our Fine Fungal Friends

Roots Demystified and *Plant Roots* were resources, as well as the aforementioned textbooks. "A. B. Frank and Mycorrhizae: The Challenge to Evolutionary and Ecologic Theory," by James Trappe, provides more detailed information about Frank's remarkable contribution.

Arsenic and Young Fronds

Articles in the *Washington Post* recount the story of the contamination and remediation of Spring Valley. Articles by Rufus Chaney et al., Lena Q. Ma et al., Agely et al., and updates by the U.S. Department of Agriculture's Agricultural Research Service and the Environmental Protection Agency provide the scientific side of the story. On *Alyssum,* see Brooks and Radford, news articles in the *Illinois Valley Daily View,* minutes from the Oregon State Weed Board, a letter from the Nature Conservancy, and the report in *Minerals Engineering Online.* Textbooks and Epstein's article explain transport in roots.

The Once and Future Wheat

My conversations with Loren Reiseberg, Professor of Botany

at the University of British Columbia and Distinguished Professor of Biology at Indiana University, as supplemented by conversations with David Van Tassel, informed my account of the history of the sunflower. For additional information on the evolution of the sunflower, see *The Sunflower* by Charles Heiser Jr.

PART III: LEAVES

A Momentous Mint

I turned to two biographies of Joseph Priestley: Schofield's *The Enlightenment of Joseph Priestley* and *The Enlightened Joseph Priestley,* and Jackson's highly readable *A World on Fire,* which tells the story of Priestley, Lavoisier, and the discovery of oxygen. Priestley never accepted the majority of Lavoisier's "new chemistry" based on measurements of mass.

In fact, Priestley ultimately viewed himself as more of a theologian than a scientist. He moved his family to Birmingham in 1780, and although he continued to conduct experiments and defend the phlogiston theory, most of his publications were theological. His primary interest was in expounding the proposition that the early Catholic Church was corrupt and that the Reformation was incomplete. He got into serious trouble when he seemed to be calling for an overthrow of the Anglican Church. He wrote, "We [the Dissenters] are, as it were, laying gunpowder, grain by grain, under the old building of error and superstition, which a single spark may hereafter inflame, so as to produce an instantaneous explosion; in consequence of which that edifice . . . may be overturned." For this, he became known as "Gunpowder Joe." He expressed such sentiments not long after the deadly chaos of the French Revolution and the murders of the French royalty, nobility, and clergy, which made for notably impolitic timing. On the second anniversary of the fall of the Bastille, which the Birmingham Dissenters celebrated, rioters attacked the two local Dissenting churches and burned them. They then moved on to burn Dissenters' homes, including Priestley's. After a time, George III sent troops to Birmingham to contain the continuing mayhem, but there is no question that the government saw Priestley and his cohorts as a threat.

Birmingham remained unsafe for Priestley, and he moved to Middlesex, where he lectured at a Dissenting academy until 1794. He became increasingly outspoken about his conviction that the new millennium would bring the Second Coming of Christ and the end of the established church. The French Revolution, in his view, had been a mere harbinger. Life in England became untenable for him: He was burned in effigy and featured in scathing political cartoons. In 1794, the Priestleys emigrated to Northumberland, Pennsylvania. The University of Pennsylvania invited him to teach chemistry, but he declined. He helped found the First Unitarian Church of Philadelphia and an academy, formed a friendship with Thomas Jefferson, and managed to stir up political controversy again when his correspondence with a radical French printer was published. In 1804, illness and sorrow over the deaths of his son Henry and his beloved wife finally dampened his effervescent spirit. He was buried in Northumberland.

Leaves Eat Air

Geerdt Magiels's *From Sunlight to Insight* is the most comprehensive, if awkwardly written, source of information on Ingen-Housz, and includes a number of his subject's letters in translation. Howard Reed's booklet, *Jan Ingenhousz, Plant Physiologist, With a History of the Discovery of Photosynthesis,* is also worth finding.

After publication of *Experiments upon Vegetables, Discovering Their great Power of purifying the Common Air in the Sun-shine, and of Injuring it in the Shade and at Night* in 1779, Ingen-Housz continued to experiment and kept up a running debate with Senebier and others on whether plants really do produce bad air at night. While he was correct on this matter, he continued to believe that the "green matter" in vials was animal in nature since it had neither seeds nor roots. He engaged with Priestley, Senebier, Spallanzani, and others over the matter of who deserved credit for discovering the role of plants in "curing" air.

Thanks to his pension, he could follow the scientific developments of the time, visiting with Réamur, the Herschels, Franklin, Coulomb, Guillotin (who did not invent the machine and actually opposed the death penalty), considering Cassini's and Halley's calculations of the distance between the sun and Earth, trying new experiments with seeds and soil, and reading about Lavoisier's experiments. A pioneer in data-based assessments of natural phenomena, he ran tests to debunk some of the strange new medical treatments of the era. One of his targets was Franz Anton Mesmer, a German physician who had gained renown for allegedly curing patients by manipulating energy forces he believed flowed through the body, either with magnets or his own "magnetic effluvium."

After the French Revolution, he lived for the most part in England, where he was often a guest on the estates of noble families. Although of modest background, he came to see society through the eyes of a man intimate with Austrian and English nobility. Shortly after the second anniversary of the Revolution, he wrote to his friend Jacob van Breda that in England the "Dissenters had attempted to overthrow church and state." Priestley, he reported, had nearly escaped a certain death, and all his writings and instruments were destroyed, "which was a terrible loss for everybody." Nonetheless, Ingen-Housz wrote, "it was a pity that such a great scientist was so cursed by fanaticism." He later wrote that Priestley is "full of pride and lust for fame."

The last years of Ingen-Housz's life were difficult. The European nations had been in a constant state of war following the revolution of 1789 and the rise of Napoleon. In 1794, Austria lost its territory in the Netherlands. The Austrian government, pressed for funds for the continuing wars, cut Ingen-Housz's pension. Meanwhile, the economic impact of the political tumult meant that he also lost his investments in various European bonds and enterprises. Plagued by kidney stones and other illnesses and unable to return to either the Netherlands or Austria, he died in September 1799, and was buried in Calne, England.

He was preceded in death by Lavoisier, whose inherited wealth and his service to the government of Louis XVI as a tax collector sent him to the guillotine during the Reign of Terror in 1794.

The Vegetable Slug

There are many books on the mechanics of photosynthesis, which is a fascinating subject that I relate only at its most simple level. I found in David Walker's *Energy, Plants, and Man* a lovely balance of good writing, scientific clarity, and clever cartoons. For the history of oxygen on Earth and the significance of photosynthesis, try Oliver Morton's *Eating the Sun* and Nick Lane's *Oxygen*.

Once in a Blue-Green Moon

The history of early Earth is, not surprisingly, the subject of much speculation and controversy. Nick Lane's *Life Ascending* and Andrew Knoll's *Life on a Young Planet* are highly readable, as is, again, Lane's *Oxygen*.

On the history of plant life through the Jurassic, I relied on *Paleobotany*, by Taylor et al.; *Fossil Plants*, by Kenrick and Davis; *The Evolution of Plants*, by Willis and McElwain; *The Evolutionary Biology of Plants*, by Niklas; and *A Natural History of Conifers*, by Farjon. See also Claire Humphreys et al., "Mutualistic Mycorrhiza-like Symbiosis . . ."

The Tenacity of Trees

I found stories of the conflicts over *Leylandii* in a number of British newspapers and journals. See "Leyland Cypress—X Cupressocyparis leylandii" at the Royal Forest Society website for the history of the hybrid.

There are many claimants to the title of world's oldest tree. The National Geographic Society is my source for the Swedish spruce.

On the history and biology of trees, both Tudge's *The*

Tree and Wilson's *The Growing Tree* are enlightening and good reads.

Amazing Grass
For background on *Miscanthus,* my sources include: *Miscanthus* for Renewable Energy Generation," by Heaton et al.; "Pros and Cons of Miscanthus," University of Illinois; "The Impact of Extensive Planting of Miscanthus . . ."; "Growth and Agronomy of *Miscanthus x giganteus* for Biomass Production," by Anderson et al. The last has an excellent bibliography.

For the evolutionary history of C4 grasses, see Kenrick and Davis, Willis and McElwain, Niklas, and the article by Edwards et al. The story is not yet fully understood: see "Biologist Solves Mystery of Tropical Grasses' Origin," Brown University; "Study Rewrites the Evolutionary History of C4 Grasses," University of Illinois; and Osborne and Beerling's "Nature's Green Revolution: The Remarkable Evolutionary Rise of C4 Plants."

According to "A Novel Mechanism by which Silica Defends Grasses Against Herbivory," by Hunt et al., the silica in *Miscanthus* discourages predators not only by making the leaves unpleasant to chew, but also by making them more difficult to digest.

PART IV: FLOWERS

Sex in the Garden
The best sources of biographical information about Sebastien Vaillant are Jacques Rousseau's article "Sebastien Vaillant: An Outstanding 18th Century Botanist" and the first chap-

ter of Roger Williams's *Botanophilia in Eighteenth-Century France: The Spirit of Enlightenment.*

Who Needs Romeo?

I relied on *Gametes & Spores: Ideas about Sexual Reproduction 1750–1914,* by John Farley, and was intrigued by Pinto-Correia's somewhat eccentric *The Ovary of Eve: Egg and Sperm and Preformation.* Olby's *Origins of Mendelism* and Roberts's *Plant Hybridization Before Mendel* are also important on the history of ideas about plant sex and inheritance. For more about Amici and the pollen tube, see the entry for Amici on the Scuola Normale Superiore website.

There is an excellent article by Claude Dolman about Spallanzani in *The Complete Dictionary of Scientific Biography,* 2008, which highlights the incredible range of the Abbe's scientific interests before and after his investigations of sex. Spallanzani's two volumes of *Dissertazioni di fisica animale e vegetabile* were printed in 1780 and 1782, and appeared in French and English a few years later. In 1779, he spent a month in Switzerland, meeting with Senebier, H. B. de Saussure (the geologist father of the botanist), Bonnet, and other naturalists. During the next five years, when he wasn't teaching at the University of Pavia he traveled extensively around Europe "observing and interrogating Nature" and collecting for the university's Museum of Natural History. He gathered specimens of hundreds of species of fish, coral, sponges, and other marine fauna from the Mediterranean and Adriatic, discovered new species of fireflies, and offered experimental proof that torpedo fish are not, as was thought, attracted by magnets.

In 1785, Spallanzani traveled to Constantinople by ship and after nearly being shipwrecked, he spent a year exploring the fauna and geology of the region. According to Dolman, "In August, 1786, having dispatched the valuable museum collections by ship, Spallanzani set out with a single attendant on the unimaginably difficult return overland. Despite hazardous mountain passes, floods and torrents, brigands and cutthroats, detours were made to inspect mines and geological structures, and more specimens were collected." Two years later, when he was nearly sixty, he traveled to southern Italy to gather information on a series of recent and highly destructive volcanic eruptions. While on Etna, the mountain erupted and toxic gases knocked him unconscious. (Shades of Pliny, who died of suffocation in A.D. 79 while observing an eruption of Vesuvius from a ship in the bay.) He ventured into the crater of Vulcano to investigate mineral structures, and emerged with burnt feet and his staff on fire. Despite the dangers, he was able to identify issuing gases and the mineral composition and temperatures of lava. Amazingly, he died peacefully in bed in 1799, shortly after his seventieth birthday.

The best source for information on Kölreuter (whose name is also spelled Koelreuter and Kolreuter) is Ernst Mayr's "Joseph Gottlieb Kölreuter's Contributions to Biology."

Black Petunias
Petunia: Evolutionary, Developmental, and Physiological Genetics was indispensable to my understanding of the petunia. Also see "Flower Development in Petunia," by van der Krol and Chua; and "Isolation Barriers . . . ," by Dell'Olivo

et al. For flower color, see "Biochemistry and Genetics of Flower Color," by Griesbach and Janick.

The Abominable Mystery

For the best photographic image of amborella online, see http://www.phytoimages.siu.edu/imgs/paraman1/r/Ambo rellaceae_Amborella_trichopoda_23986.html. Pam Soltis, Doug Soltis, and their colleagues have written extensively about amborella and the evolution of angiosperms: see "The Amborella genome . . . ," "Floral Developmental Morphology . . . ," "The Making of the Flower," and "The floral Genome: An evolutionary History," among others.

On the evolution of flowering plants and their unique anatomy, I turned to *Floral Biology: Studies on Floral Evolution in Animal-Pollinated Plants* and a number of more technical articles. A relatively simple explanation of the role of LEAFY and other genes can be found in "Age-Old Question on Evolution of Flowers Answered." More detailed is "A Short History of MADS-Box Genes in Plants." For the evolution of endosperm, see "The Evolutionary Origins of the Endosperm in Flowering Plants." For the evolution of nectar, see "Nectar: Properties, Floral Aspects, and Speculations on Origin." For the history of seeds, Ada Linkies and her colleagues have a clear exposition in "The Evolution of Seeds." I also found "After a Dozen Years of Progress the Origin of Angiosperms Is Still a Great Mystery" enlightening. The Wikipedia entry "Evolutionary History of Plants" has an extensive list of helpful sources.

Cheap Sex

Stefan Vogel's chapter on Sprengel in *Floral Biology* is essential.

There are many books on Darwin, of course, but I primarily relied on *Darwin: The Life of a Tormented Evolutionist,* by Desmond and Moore; *Darwin, His Daughter & Human Evolution,* by Keynes; and especially the wonderful *The Aliveness of Plants: The Darwins at the Dawn of Plant Science,* by Ayres. "Darwin's Botany," by Ornduff, was also helpful, as was Oliver Sacks's "Darwin and the Meaning of Flowers."

On orchids and mimicry, *An Enthusiasm of Orchids* by John Alcock is a delight. For the tidbit on the *Ophrys sphegodes,* see Schiestl in *Oecologia.*

PART V: ONWARD, UPWARD, AND AFTERWARD

Onward and Upward

Darwin's books *The Movements and Habits of Climbing Plants* and *The Power of Movement in Plants* are very readable. For a more complete history of the discovery of auxin post-Darwin, see "The Odyssey of Auxin," by Abel and Theologis.

Afterward

The European Space Agency estimates that there 10^{22} stars in the universe. Given that there are about 500,000 chloroplasts in a square millimeter of leaf (http://hyperphysics. phy-astr.gsu.edu/hbase/biology/chloroplast.html, among other sources) and using the estimate of 8,000 square feet of foliage on a sixty-foot oak tree, about 27 million trees hold as many chloroplasts as there are stars in the universe. Given that NASA estimates there are 400 billion trees (not to mention nonwoody plants and photosynthetic algae) on the planet (http://www.npr.org/templates/story/story. php?storyId=96758439), the number of chloroplasts on Earth is awe-inspiring.

Bibliography

Darwin, Charles. *The Movements and Habits of Climbing Plants.* 1865, printed from Web-based open source.

———. *The Power of Movement in Plants.* 1865, printed from Web-based open source.

Grew, Nehemiah. *Anatomy of Plants with An Idea of a Philosophical History of Plants and Several Other Lectures Read Before the Royal Society.* 1682; reprint, New York and London: Johnson Reprint, 1965.

Hales, Stephen. *Vegetable Staticks.* 1727; reprint, London: Macdonald; New York: American Elsevier, 1969.

Hooke, Robert. *Micrographia; or some physiological descriptions of minute bodies made by magnifying glasses with observations and inquiries thereupon.* 1665; reprint, Bruxelles: Culture et Civilisation, 1966.

Malpighi, Marcello. *The Correspondence of Marcello Malpighi 1628–1694.* Ed. Howard B. Adelmann. Ithaca, NY: Cornell University Press, 1975.

Pliny, the Elder. *Natural History.* Cambridge, MA: Harvard University Press, Loeb Classical Library, 1936.

Spallanzani, Lazzaro. *Dissertations Relative to the Natural History of Animals and Vegetables.* Translated from the Italian of the Abbe Spallanzani, to which are added two letters from Mr. Bonnet to the author, 1784. Available through Eighteenth Century Collections Online.

Theophrastus. *Inquiry Into Plants* and *De Causis Plantarum.* Translated by Arthur Hort. Cambridge, MA: Harvard University Press, Loeb Classical Library, 1936.

SECONDARY SOURCES

Abel, S. and Theologis, A. "The Odyssey of Auxin." *Cold Spring Harbor Perspectives in Biology* 2, no. 10 (October 2010, http://www.ncbi.nlm.nih.gov/pubmed/20739413.

Agely, Abid Al, et al. "Mycorrhizae Increase Uptake by the Hyperaccumulator Chines Brake Fern (*Pteris vittata L.*)." *Journal of Environmental Quality,* published online November 5, 2005, www.ncbi.nlm.nih.gov/pubmed/16275719.

Alcock, John. *Orchids: Sex and Deception in Plant Evolution.* Oxford: Oxford University Press, 2006.

Allan, D. G. C., and Robert E. Schofield. *Stephen Hales: Scientist and Philanthropist.* London: Scolar Press, 1980.

Anderson, Eric, et al. "Growth and Agronomy of Miscanthus x giganteus for Biomass Production." *Biofuels* 2, no. 2 (March 2011): 167–83.

Arber, Agnes Roberson. "Nehemiah Grew and Marcello Malpighi." *Proceedings of the Linnaean Society of London* 153, no. 2 (November 1941): 218–38.

——. "Nehemiah Grew (1641–1712) and Marcello Malpighi (1628–1694): An Essay in Comparison." *Isis* 34, no. 1 (1942–43): 7–16.

——. "Nehemiah Grew and the Study of Plant Anatomy." In *Science Progress in the Twentieth Century: A Quarterly Journal of Scientific Work & Thought,* vol. 1, 1906. Available as a Google e-book.

"Arsenic, Illnesses Worry D.C.; Unusual Ailments Near Tainted Sites." *Washington Post,* January 27, 2001.

Ayres, Peter. *The Aliveness of Plants: The Darwins at the Dawn of Plant Science.* London and Brookfield, VT: Pickering & Chatto, 2008.

Baroux, Célia, et al. "The Evolutionary Origins of the Endosperm in Flowering Plants." *Genome Biology* 3, no. 9 (2002): reviews1026.1–reviews1026.5.

Berg, Linda R. *Introductory Botany: Plants, People and the Environment.* Pacific Grove, CA: Brooks-Cole—Thomson Learning, 1997.

Birken, William. "The Dissenting Tradition in English Medicine of the Seventeenth and Eighteenth Centuries." *Medical History* 39 (1995): 197–218.

Bolam, Jeanne. "The Botanical Works of Nehemiah Grew, F.R.S. (1641–1712)." *Notes and Records of the Royal Society of London* 27 (1973): 219–31.

Bradbury, Savile. *The Evolution of the Microscope.* Oxford and New York: Pergamon Press, 1967.

Brady, Nyle, and Ray Wei. *The Nature and Properties of Soils.* Upper Saddle River, NJ: Pearson Education, 2008.

Brooks, R .R., and C. C. Radford. "Nickel Accumulation by European Species of the Genus Alyssum." *Proceedings of the Royal Society of London B* 200 (1978): 217–24.

Brown University. "Biologist Solves Mystery of Tropical Grasses' Origin." Press release, February 9, 2010, http://news.brown .edu/pressreleases/2010/02/grasses.

Buzgo, Matyas, P. Soltis, et al. "The Making of the Flower." *Biologist* 52, no. 3 (July 2005).

Buzgo, Matyas, P. Soltis, and D. Soltis. "Floral Developmental Morphology of *Amborella Trichopoda* (Amborellaceae)." *International Journal of Plant Sciences* 165, no. 6 (November 2004): 925–47, http://www.jstor.org/stable/10.1086/424024.

Bynum, William. "The Great Chain of Being After Forty Years: An Appraisal." *History of Science* 13 (1975): 1–28.

Capon, Brian. *Botany for Gardeners.* Portland, OR: Timberland Press, 2005.

Carrubba, Robert. "Englebert Kaempfer and the Myth of the Scythian Lamb." *Classical World* 87 (1993): 417.

Chaney, R. L. "Plant Uptake of Inorganic Waste Constituents." In *Land Treatment of Hazardous Wastes,* ed. J. F. Parr, P. B. Marsh, and J. M. Kla. Park Ridge, NJ: Noyes Data, 1983, pp. 50–76.

Chaney, R. L., C. L. Broadhurst, and T. Centofanti. "Phytoremediation of Soil Trace Elements." In *Trace Elements in Soils,* ed. P. S. Hooda. Chichester, UK: John Wiley, 2010.Clark-Kennedy, A. E., *Stephen Hales D.D., F.R.S.: An Eighteenth Century Biography.* Ridgewood, NJ: Gregg Press, 1965.

Correia, Clara Pinto. *The Ovary of Eve: Egg and Sperm and Preformation.* Chicago: University of Chicago Press, 1997.

Crossland, Maurice, ed. *The Emergence of Science in Western Europe.* New York: Science History, 1976.

Dear, Peter. *Revolutionizing the Sciences: European Knowledge and Its Ambitions, 1500–1700.* Princeton: Princeton University Press, 2001.

de la Barrera, E., and P. S. Nobel. "Nectar: Properties, Floral Aspects, and Speculations on Origin." *Trends in Plant Science* 9, no. 2 (February 2004): 65–69, http://www.ncbi .nlm.nih.gov/pubmed/15102371.

Delaporte, François. *Nature's Second Kingdom: Explorations of Vegetality in the Eighteenth Century.* Trans. Arthur Goldhammer. Cambridge, MA: MIT Press, 1982.

Dell'Olivo, A., et al. "Isolation Barriers between *Petunia Axillaris* and *Petunia Integrifolia* (Solanaceae)." *Evolution* 65, no. 7 (July 2011): 1979–91.

Desmond, Adrian, and James Moore. *Darwin: The Life of a Tormented Evolutionist*. New York: Norton, 1991.

Edwards, Gerald, et al. "What Does It Take to Be C4? Lessons from the Evolution of C4 Photosynthesis." *Plant Physiology* 125 (January 2001): 46–49, www.plantphysiol.org.

Encylopedia.com. "Spallanzani." www.encylopedia.com/topic/Lazzaro_Spallanzani.aspx.

Environmental Protection Agency. Mid-Atlantic Superfund. "Washington, D.C. Army Chemical Munitions (Spring Valley) Current Site Information." http://www.epa.gov/reg3hwmd/npl/DCD983971136.htm, retrieved March 11, 2011.

Epstein, Emanuel. "Spaces, Barriers, and Ion Carriers: Ion Absorption by Plants." *American Journal of Botany* 47 (May 1960): 393–99.

Evans, R. G. *The University of Cambridge: A New History*. London and New York: I. B. Tauris; New York: Palgrave Macmillan, 2010.

Evenden, Doreen. *Popular Medicine in Seventeenth Century England*. Bowling Green, OH: Bowling Green State University Popular Press, 1988.

"The Evolution of Orchid and the Orchid Bee." *Smithsonian,* September 23, 2011, http://blogs.smithsonianmag.com/science/2011/09/the-evolution-of-the-orchid-and-the-orchid-bee.

"Excavation by Military Forces Some AU Closings; Buildings, Homes to Be Emptied for Dig; Neighbors Concerned." *Washington Post,* January 8, 2001.

Farjon, Aljos. *A Natural History of Conifers*. Portland, OR: Timber Press, 2008.

Farley, John. *Gametes & Spores: Ideas about Sexual Reproduction 1750–1914*. Baltimore: Johns Hopkins University Press, 1982.

Ferrari, Giovanni. "Public Anatomy Lessons and the Carnival: The Anatomy Theatre of Bologna." *Past & Present*, no. 117 (November 1987): 50–106.

"Fir Extinguisher." *BBC News Magazine*, May 31, 2005.

Fournier, Marion. *The Fabric of Life: Microscopy in the Seventeenth Century*. Baltimore: Johns Hopkins University Press, 1996.

Frohlich, Michael, and Mark Chase. "After a Dozen Years of Progress the Origin of Angiosperms Is Still a Great Mystery." *Nature* 450, no. 27 (December 2007): 1184–89, http://www.nature.com/nature/journal/v450/n7173/abs/ nature06393.html.

"Gardener Is Shot Dead in Hedge Feud." *Telegraph*, July 7, 2000.

Gerats, Tom, and Judy Strommer. *Petunia: Evolutionary, Developmental and Physiological Genetics*. New York and London: Springer, 2008.

Griesbach, R. J. and Jules Janick. *Plant Breeding Reviews*, vol. 25, "Chapter 4: Biochemistry and Genetics of Flower Color." Published online June 22, 2010.

Heaton, Emily, et al. "Miscanthus for Renewable Energy Generation: European Union Experience and Projections for Illinois." *Mitigation and Adaption Strategies for Global Change* 9, no. 4 (October 2004): 433–51.

Heiser, Charles. *The Sunflower*. Norman: University of Oklahoma Press, 1976.

Humphreys, Claire, et al. "Mutualistic Mycorrhiza-like Symbiosis in the Most Ancient Group of Land Plants." *Nature Communications* 1 (November 2010): 103.

Hunt, J. W., et al. "A Novel Mechanism by which Silica Defends Grasses Against Herbivory." *Annals of Botany* 102, no. 4 (October 2008): 653–56, published online August 11, 2008.

Hunter, Michael C. W. *Establishing the New Science: The Experience of the Early Royal Society.* Woodbridge, UK, and Wolfeboro, NH: Boydell Press, 1989.

Illinois Valley Daily. "Durable Alyssum Spreads Beyond Welcome." March 1, 2001, http://www.ivdailyview .com/2011/03/01/durable-plant-spreads-beyond-welcome/, retrieved March 29, 2011.

Inwood, Stephen. *Forgotten Genius: The Biography of Robert Hooke.* San Francisco: MacAdam/Cage, 2003.

Jackson, Joe. *A World on Fire: A Heretic, an Aristocrat, and the Race to Discover Oxygen.* New York: Viking, 1977.

Jardine, Lisa. *The Curious Life of Robert Hooke: The Man Who Measured London.* New York: HarperCollins, 2004.

Kenrick, Paul, and Paul Davis. *Fossil Plants.* London: Natural History Museum, 2004.

Keynes, Randal. *Darwin, His Daughter and Human Evolution.* New York: Riverhead Books, 2001.

Knoll, Andrew H. *Life on a Young Planet: The First Three Billion Years of Evolution on Earth.* Princeton: Princeton University Press, 2005.

Kourik, Robert. *Roots Demystified.* Occidental, CA: Metamorphic Press, 2008.

Lane, Nick. *Life Ascending: The Ten Great Inventions of Evolution.* New York : Norton, 2009.

——. *Oxygen: The Molecule That Made the World.* Oxford: Oxford University Press, 2009.

Lee, Henry. *The Vegetable Lamb of Tartary; A Curious Fable of the Cotton Plant to which is Added a Sketch of the History of Cotton and the Cotton Trade.* London: Sampson Low, Marston, Searle, & Rivington, 1887.

LeFanu, William R. "The Versatile Nehemiah Grew." *Proceedings of the American Philosophical Society* 115, no. 6 (December 30, 1971): 502–506.

Letter from the Nature Conservancy to the Oregon State Weed
Board, attn: Tim Butler, dated February 10, 2009, and
summing up the "substantial threat to the native serpentine
flora of the Illinois Valley."

"Leyland Cypress—X Cupressocyparis leylandii." Royal Forest
Society, www.rfs.org.uk/learning/leyland-cypress.

"Leylandii dispute ends in light relief," *The Telegraph*, May 17,
2008.

Linkies, Ada, et al. "The Evolution of Seeds." *New Phytologist*
186, no. 4 (June 2010): 817–31, http://www.ncbi.nlm.nih.gov/
pubmed/20406407.

Lloyd, David, and Spencer Barrett. *Floral Biology: Studies on
Floral Evolution in Animal-Pollinated Plants.* New York:
Chapman & Hall, 1996.

Lovejoy, Arthur O. *The Great Chain of Being: A Study of the
History of an Idea.* New York: Harper & Row, 1936, 1960.

Ma, Lena Q, et al. "A Fern That Hyperaccumulates Arsenic."
Nature 409 (February 2001).

Magiels, Geerdt. *From Sunlight to Insight: Jan IngenHousz,
the Discovery of Photosynthesis and Science in the Light of
Ecology.* Brussels: Uitgeverilj VUBPRESS, 2010.

Mayr, Ernst. "Joseph Gottlieb Kölreuter's Contributions to
Biology." *Osiris*, 2nd series (1986): 135–76.

McCormick, J. B., and Gerard L'E. Turner. *The Atlas Catalogue
of Replica Rara Ltd. Antique Microscopes (1675–1840).*
Chicago and London: Replica Rara, 1975.

Meli, Domenico, ed. *Marcello Malpighi, Anatomist and
Physician.* Firenze: Leo S. Olschki, 1997.

Meli, Domenico. *Mechanism, Experiment, Disease.* Baltimore:
Johns Hopkins University Press, 2011.

———. "Mechanistic Pathology and Therapy in the Medical
Assayer of Marcello Malpighi." *Medical History* 51, no. 2
(April 2007): 165–80.

Minerals Engineering International Online. "Green Nickel."
    November 22, 2002, http://www.min-eng.com/commodities/
    metallic/nickel/news/1.html, retrieved March 31, 2011.

Morton, A. G. *The History of Botanical Science: An Account
    of the Development of Botany from Ancient Times to the
    Present Day*. London and New York: Academic Press, 1981.

Morton, Oliver. *Eating the Sun: How Plants Power the Planet*.
    New York: Harper Perennial, 2009.

"Mother of All Trees That Sets Neighbours at War . . ." *Western
    Mail*, January 26, 2008.

National Geographic News. "Oldest Living Tree Found in
    Sweden." April 14, 2008, http://news.nationalgeographic.
    com/news/2008/04/080414-oldest-tree.html.

National Science Foundation. "What 'Pine' Cones Reveal About
    the Evolution of Flowers." *ScienceDaily*, Dec. 14, 2010, online
    Feb. 6, 2013.

Nicolson, Marjorie Hope. *Pepys' Diary and the New Science*.
    Charlottesville: University Press of Virginia, 1965.

Niklas, Karl J. *The Evolutionary Biology of Plants*. Chicago:
    Chicago University Press, 1997.

Northcoast Environmental Center. "Rare Habitat Threatened
    by Imported Weeds." February 2010, http://yournec.org/
    content/rare-habitat-threatened-imported-weeds, retrieved
    April 1, 2011.

Olby, Robert, *Origins of Mendelism*. New York, Schocken
    Books, 1967.

Oregon State Weed Board. Minutes, February 19-20, 2009, www
    .oregon.gov/ODA/PLANT/ . . . /minutes_salem_09.pdf.

Ornduff, Robert. "Darwin's Botany." *Taxon* 33, no. 1 (February
    1984): 39–47.

Osborne, Colin, and David Beerling. "Nature's Green
    Revolution: The Remarkable Evolutionary Rise of C_4 Plants."
    Philosophical Transactions of the Royal Society of London

B, Biological Sciences 361, no. 1465 (January 2006): 173–94, published online Nov. 28, 2005.

Pirozynski, K. A., and D. W. Malloch. "The Origin of Land Plants: A Matter of Mycotrophism." *Biosystems* 6, no. 3 (March 1975): 153–64.

Pollan, Michael, "Love and Lies." *National Geographic,* September 9, 2009.

Reed, Howard, and Jan Ingenhousz. *Plant Physiologist, With a History of the Discovery of Photosynthesis.* Reprinted from the *Chronica Botanica* 11, no. 5/6. Waltham, MA: Chronica Botanica; New York: Stechert-Hafner, 1950, pp. 285–396.

Roberts, H. F. *Plant Hybridization Before Mendel.* 1929; reprint, New York: Hafner, 1965.

Rousseau, Jacques. "Sébastien Vaillant, an Outstanding 18th-Century Botanist," *Regnum vegetabile* 71 (1970): 195–228.

Sachs, Julius von. *History of Botany (1530–1860).* Trans. Henry E. F. Garnsey, rev. Isaac Bayley Balfour. 1890; reprint, New York: Russell & Russell, 1967.

Sacks, Oliver. "Darwin and the Meaning of Flowers." *New York Review of Books,* November 20, 2008.

Schiestl, F., and M. Ayasse. "Post-pollination Emission of a Repellent Compound in a Sexually Deceptive Orchid: A New Mechanism for Maximising Reproductive Success?" *Oecologia* 126, no. 4 (2001): 531–34.

Schofield, Robert E. *The Enlightened Joseph Priestley: A Study of His Life and Work 1773–1804.* University Park: Pennsylvania State University Press, 2004.

———. *The Enlightenment of Joseph Priestley: A Study of His Life and Work 1733–1773.* University Park: Pennsylvania State University Press, 1997.

Scott, Peter. *Physiology and Behaviour of Plants.* Chichester, UK, and Hoboken, NJ: John Wiley, 2008.

Scuola Normale Superiore. "Amici" entry, http://gbamici.sns.it/eng/osservazioni/osservazionibiologiche.htm.

Soltis, D., et al. "The Floral Genome: An Evolutionary History of Gene Duplication and Shifting Patterns of Gene Expression." *Trends in Plant Science* 12, no. (August 2007): 358–67, http://www.ncbi.nlm.nih.gov/pubmed/17658290.

Soltis, Douglas, et al. "The *Amborella* Genome: An Evolutionary Reference for Plant Biology," *Genome Biology* 9, no. 402 (2008), http://genomebiology.com/2008/9/3/402.

Theissen, Gunter, et al. "A Short History of MADS-box Genes in Plants." *Plant Molecular Biology* 42 (2000): 115–49, http://www.ncbi.nlm.nih.gov/pubmed/10688133.

Trappe, James. "A. B. Frank and Mycorrhizae: The Challenge to Evolutionary and Ecologic Theory." *Mycorrhiza* 15, no. 4 (June 2005): 277–81.

Tudge, Collin. *The Tree: A Natural History of What Trees Are, How They Live, and Why They Matter.* New York: Crown, 2006.

Twigg, John. *The University of Cambridge and the English Revolution, 1625–1688.* Woodbridge, UK, and Rochester, NY: Boydell Press; Cambridge: Cambridge University Library, 1990.

Unisci. "Age-old Question on Evolution of Flowers Answered." http://www.unisci.com/stories/20012/0615015.htm?iframe=true&width=100%&height=100%. June 15, 2001.

University of Illinois at Urbana-Champaign. "Study Rewrites the Evolutionary History of C4 Grasses." Press release, November 16, 2010, news.illinois.edu/news/10/1116paleoecology.html.

University of Illinois. "Pros and Cons of Miscanthus." Press release, September 14, 2010, http://web.extension.illinois.edu/state/newsdetail.cfm?NewsID=18968.

U.S. Department of Agriculture, Agricultural Research Service. "Using Plants to Clean Up Soils." January 29, 2007.

Waisel, Yoav, Amram Eshel, and Uzi Kafkafi, eds. *Plant Roots: The Hidden Half.* New York: Marcel Dekker, 1991.

Walker, David. *Energy, Plants and Man*. East Sussex, UK: Oxygraphics, 1992.

Williams, Roger. *Botanophilia in Eighteenth-Century France: The Spirit of Enlightenment*. Dordrecht, Netherlands: Kluwer Academic, 2001.

Willis, K. J., and J. C. McElwain. *The Evolution of Plants*. New York: Oxford University Press, 2002.

Wilson, Brayton. *The Growing Tree*. Amherst: University of Massachusetts Press, 1970.

Wilson, Catherine. *The Invisible World: Early Modern Philosophy and the Invention of the Microscope*. Princeton: Princeton University Press, 1995.

"WWI Munitions Unearthed at D.C. Construction Site." *Washington Post*, January 6, 1993.

Index